Lilia BOUSSOUF
Meriem CHABOUNIA
Nesrine DJAFRI

Impact of drying on the quality of figs (Ficus carica)

Lilia BOUSSOUF
Meriem CHABOUNIA
Nesrine DJAFRI

Impact of drying on the quality of figs (Ficus carica)

Impact of drying on the physicochemical properties, nutritional quality and antioxidant activity of figs (Ficus carica L).

ScienciaScripts

Imprint
Any brand names and product names mentioned in this book are subject to trademark, brand or patent protection and are trademarks or registered trademarks of their respective holders. The use of brand names, product names, common names, trade names, product descriptions etc. even without a particular marking in this work is in no way to be construed to mean that such names may be regarded as unrestricted in respect of trademark and brand protection legislation and could thus be used by anyone.

Cover image: www.ingimage.com

This book is a translation from the original published under ISBN 978-620-3-45690-5.

Publisher:
Sciencia Scripts
is a trademark of
Dodo Books Indian Ocean Ltd. and OmniScriptum S.R.L publishing group

120 High Road, East Finchley, London, N2 9ED, United Kingdom
Str. Armeneasca 28/1, office 1, Chisinau MD-2012, Republic of Moldova, Europe
Printed at: see last page
ISBN: 978-620-5-97830-6

ACKNOWLEDGEMENTS

In the name of Allah the Most Gracious, Praise be to God, the Only One, the Almighty who has He gave us strength, health and patience and guided our steps to accomplish this modest work. His faith alone has inspired and comforted us to reach this goal. We address our sincere thanks to our promoter, Mrs. BOUSSOUF Lilia, Doctor in the department of Applied Microbiology and Food Sciences, Faculty of Natural and Life Sciences, Mohamed Seddik Benyahia University of Jijel, for having accepted the scientific direction of our end of cycle project. May she be reassured of our deep gratitude. We are very honored to thank also Mrs. BEKKA Fahima, Doctor, at the University Mohamed Seddik Benyahia of Jijel for the great privilege that she made us to have accepted the presidency of our jury of defense. May she be assured of our respectful consideration. We also thank Mrs. DJABALI Saliha, Maitre Assistante A at the Université Mohamed Seddik Benyahia de Jijel, for having accepted to devote time to examine and judge this work. A big thank you to our teachers, Mr. BOUBZARI Mohammed Tahar and Miss. LABSIR Dalila for all the advice and encouragement they gave us.

DEDICATION

The studies are first and foremost our only asset Wishing that the fruit of our efforts day and night will lead us to blossoming happiness.At the time of the assessment, I cannot take credit for this work alone. And it is therefore logical that I would like to thank the people who contributed to it, so at the end of this humble researchFirst and foremost to two of the dearest people in the world who are burning for their daughter to see the light of knowledge, who have made many sacrifices, who have raised me, trained me, encouraged me and supported me throughout my life, and who continue to ensure my success in my studies: My mother Malika and my father Mourad whom I have loved so much with great affection and I am very proud to have them and all the words in the world cannot express the love and respect that I have for them May God protect you, a life language Inch-Allah.I dedicate this modest work to my dear husband for his support, his patience and his encouragement as well as to my precious brothers Badis and Mohammed to whom I wish all happiness and success without forgetting my dear cousins Ahlem and Selma for their precious help. My thanks also go to my dear friend and partner Nesrine as well as her family for all her support, her kindness and her seriousness which allowed us to accomplish this work, I wish her all the success and happiness in the world and that God grants all these wishes.

Meriem

CONTENTS

INTRODUCTION

Fruit growing is an integral part of the economic and social life of Algeria (Benettayeb, 1993). The happy harmonization of the fruit vocation of this country, which results from its climate, its possibilities of irrigation and the development of the consumption of fruits as well in Algeria as on the foreign markets led him to put in culture several fruit species.The fig tree belongs with the olive tree and the vine to the trilogy of the main fruit productions of Algeria. This importance is mainly related to a multiplicity of uses and exchanges of genetic material which led to its diversification and propagation (INRAA, 2006).The fig is the fruit of the fig tree, a majestic tree of the family Moraceae, which is the emblem of the Mediterranean basin, where its culture and its use constitute an ancient tradition. This thousand-year-old fruit has accompanied, nourished and given much pleasure to our ancestors. Due to its nutritional potential, the fig has been a staple food of the Mediterranean populations since antiquity, as well as a dietetic product.The inhabitants of the northern regions of the country are very high consumers of figs. These are seasonal and very perishable due to their short shelf life which is two days at room temperature (Sharifian et al., 2012), and 7 to 10 days if stored between 0 and 2°C (Veberic et al., 2008). In order to extend their shelf life and meet the demands of the 21st century consumers several types of processing are applied for this fruit, but still the most used process is drying (Vidaud, 1997).The drying technique is an early technique but it is still widely used nowadays (Eshak, 2018). Its main objective is to increase the shelf life of the product by reducing the moisture content and water activity in order to inhibit microbial growth and enzyme activity (Aguilera et al., 2003; Cano-Chauca et al., 2004). Once dried, figs can be stored for 6-8 months (Slatnar et al., 2011). Drying also results in a substantial reduction in weight and volume, minimizing packaging, storage and transportation costs (Okos, 1992).Indeed, the drying can be achieved by several processes, among them we cite the sun drying, the latter is used for a long time by farmers and produces only a small amount of dry figs (El Khaloui, 2010). It requires little capital, a simple equipment and low energy input (Hoxha and Kongoli, 2016). Nevertheless, the slow process, which is dependent on climatic conditions, can lead to various problems such as growth and proliferation of microorganisms and loss of nutritional compounds.Another commonly used method is oven drying (hot air dehydration) which has gained importance due to its many advantages over sun drying. Indeed, this process is faster, is done in better sanitary conditions, and its parameters can be easily controlled, nevertheless it is much more energy consuming (Slatnar et al., 2011).To reduce the aforementioned problems, new techniques have been studied, such as those assisted by microwave radiation. These are promising alternatives for drying fruits and vegetables, reducing time and energy consumption and improving the sensory and technological quality of the dried pulp (Ferrao et al ., 2017).We chose the dried fig because on the one hand it has great food, economic and heritage importance, and on the other hand its consumption manifests a positive effect on human health, thanks to the energy it provides through its high sugar content and its richness in fiber, antioxidant polyphenols as well as other compounds with appreciable amounts such as minerals, proteins, organic acids and volatile compounds that provide a characteristic pleasant aroma (Solomon et al., 2006; Oliveira et al., 2009; Bachir Bey et al., 2013).Recently, many investigations have focused on the effect of different drying processes on the physicochemical properties of this fruit, as well as on its nutritional and pharmacological quality.In this perspective we are interested in presenting a bibliographical study on the impact of drying on the physicochemical properties, nutritional quality, phytochemical and especially on the antioxidant activity of the fig (Ficus carica L.). For that our thesis of end of cycle will be articulated as follows:

• The first chapter deals with the agronomic aspect of the fig tree.
• The second and third chapters will include bibliographic data on figs and drying, respectively.

• The fourth chapter will be devoted to previous studies dealing with the influence of different drying processes, namely sun, oven and microwave drying, on the quality attributes of the fig.

CHAPTER I
THE AGRONOMIC ASPECT OF THE FIG TREE FICUS CARICA L.

I.1. General information on the fig tree

Ficus carica Linn. or more commonly known as figue, is one of the earliest cultivated fruits in human history. Today, it is consumed worldwide (Mat Desa et al., 2019)."Ficus carica L." comes from the word "Ficus" meaning wart, as its milk cures this pathology, and the word "carica" indicating a region in Turkey where it probably first existed (Vidaud, 1997).In polytheistic and monotheistic religions, the fig tree has a sacred aspect (El Bouzidi, 2002), the fig is also the most frequently mentioned fruit in the Bible and was cited in the "Surat Attine" of the Holy Qur'an: "By the fig tree and the olive tree" (Qur'an 95:1), which contributed to its sacredness in Muslim society.Because of its antiquity, the origin of F. carica is subject to divergent hypotheses, but all admit that it precedes the domestication of wheat (Benettayeb, 2018). According to Jeddi, (2009) the fig tree was known in the Middle East since the third millennium among the ancestors of the Sumerians, this species was cultivated by the Phoenicians, Egyptians and Greeks to the point where it is thought to be a plant indigenous to these environments.Remains of fruit dating back to 7000 BC have been found in the Jericho excavations (Flaishman et al., 2008). Charles, (2002) specified that it was the Carthaginians and then the Greeks who extended the cultivation of the common fig tree throughout the Mediterranean basin. The fig tree has thus gradually spread in culture and reached the most distant countries.

I.2. Classification botanical

The common fig (F. carica; 2n = 26) belongs to the family Moraceae which has about 1500 species in 40 genera. The genus Ficus, which is usually classified into six sections or subgenera (Ficus, Synoecia, Sycidium, Sycomorus, Pharmacosycea, Urostigma) includes 700 to 800 species. However, only the species F. carica and F. sycomorus (Egyptian sycamore fig) have edible fruits (Berg and Wiebes, 1992).

I.2.1. Systematic position of the fig tree

Systematically, the taxonomy of fig tree reported by Chawla et al., (2012) is as follows:

Table 01: Taxonomy of fig tree (Chawla et al., 2012).

Kingdom	Plant
Super-tied	Spermatophyte
Branch	Phanerogams
Class	Dicotyledonous
Subclass	Hamamelidae
Order	Urticale
Family	Moraceae
Type	Ficus
Species	Ficus carica

The classification of Ficus taxa divides fig trees into four horticultural forms, namely the wild or caprifuge type and the cultivated or domesticated forms (Tous and Ferguson, 1996) :

• The caprifiguier : Does not give edible fruits (no evolution of the flowers into fruits), but hosts blastophages which ensure the caprification of the figs.

• The domestic fig tree: According to Condit's monograph (1995), there are three types of domestic fig trees: Smyrna (require pollination), San Pedro (flowering figs do not require pollination unlike figs) and Common (figs and flowering figs do not require pollination). Of which 78% of the fig varieties in the world are of the Common type, less than 4% are of the San Pedro type and the remaining 18% are of the Smyrna type. The prevalence of the Common type in the world has probably been favored by selection because of its parthenocarpy, especially in regions without blastophagy.

Similarly, the fig tree has a great adaptability and an amazing capacity of vegetative regeneration and fruit production without visible flowers.

The production of this species is of two types (Oukabli, 2003):

• Bifacial figs (flower figs): The figs of first harvest or flower figs (El Bacor) are formed on the leafless branches of the previous year. They spend the winter at the peppercorn stage to resume their development in spring. The evolution of the figs flowers does not require pollination and is done in a parthenocarpic way.

• Unified (Fall) Figs: Figs are formed in the leaf axils of growing twigs.

A time lag of a few weeks is always observed between the ripening times of these two types of production (Oukabli, 2003).

I.3. Biology of the fig tree

I.3.1. Description and morphology of the fig tree

The fig tree is a bulky, vigorous and long-lived tree. It is usually grown as a shrub of 2-5 m in height, but in free-running form it can exceed 10 or 12 m. The vegetative constitution of the tree is semi-woody (Benettayeb' 2018). The latex of the plant is milky white contains mainly ficin (a protein digesting enzyme) (Chawla et al.,2012).

I.3.1.1. Trunk and bark

Its trunk is twisted, stubby and swollen at the nodes (Benettayeb' 2018). The bark is smooth with few cracks of a light gray color, retaining for a long time the traces of leaf insertion and the characteristic annular scar left by the stipules. This bark occurs on 2-3 year old parts, with younger parts changing from a soft green epidermis to a glazed brown ornamented with numerous large lenticels (Vidaud, 1997).

I.3.1.2. Fruiting shoots

The shoot is made up of a set of nodes, each node constitutes the insertion point of a leaf and axillary buds (Figure 01). Their alternate arrangement, rarely opposite on the shoot, is a specificity of the Moraceae family (Vidaud, 1997).

Figure 01: Young branch of the year (Vidaud, 1997)

I.3.1.3. Bud terminal

The fig tree consists of a terminal bud (Figure 02). The latter is made up of two stipules corresponding to the last leaf put in place. In this bud are 9 to 11 leaf outlines with their stipules (Vidaud, 1997).

Figure 02 : Photo of a terminal bud of fig tree (Vidaud ,1997)

I.3.1.4. Sheets

Fig leaves are green, alternate, palmate, odorless (Chawla et al., 2012), broad, oval or orbicular (Patil and Patil, 2011), usually having 3 to 5 lobes (Joseph and Justin, 2011) (Figure 03). The upper surface is rough and the lower surface pubescent, with veins that contribute raw sap and receive elaborated sap (Chawla et al., 2012).

Figure 03: Shape of the fig tree leaf (Vidaud, 1997)

I.3.1.5. Inflorescence and flowers

The inflorescence of the fig tree is very particular. The flowers are not visible outside, they are enclosed in a kind of receptacle called sycone which has an opening, the ostiole, which opens opposite the short peduncle carrying the figs (Vidaud, 1997). The flowers found in figs can be of two types, either male containing both female and male flowers, the latter being few in number and located all around the ostiole, or female consisting only of female flowers with a long style (Vidaud, 1997).

I.3.1.6. Root system

It is usually shallow and spreading, sometimes covering 50 feet of soil, but in permeable soil, some roots can reach down to 20 feet. Many Ficus plants produce aerial roots that descend to the ground. In some species native to tropical forests, the small Ficus plant takes up residence at the top of a tree and, by dropping aerial roots, gradually overcomes and strangles its host (Chawla et al., 2012).

I.3.2. development cycle

Ficus species are gynodioic, and functionally dioecious. Some are functionally female and produce only a seed fruit, while others are functionally male and produce only pollen and wasp pollen progeny. (carrier pollen), it is then Blastophaga psenes that brings the pollen from the male flower to the female flower (Janzen, 1979; Wiebes, 1979; Kjellberg et al., 1987).

I.3.2.1. Physiology of the development cycle

The fig tree, is a firm receptacle, an urn called synconium or syconium. The flowers are not visible, to see them you have to open the fig. Because of this shape, the inflorescence represents a mechanical barrier for the dispersion of pollen; this barrier is lifted thanks to the intervention of the pollinating insect, the blastophagus (Figure 04).The description of the biological cycle (Figure 05) begins in winter, when the fig and the insect (cycle 1a, 1b) are at rest. The cycle does not resume until April with the establishment of a new fig shoot (cycle 2a, 2b) and the resumption of the development of the blastophagus larvae (cycle 2a), from which the adult female emerges in May without being loaded with pollen because the male flowers of the caprifuge have no pollen (cycle 3a) (Kjellberg et al., 1987).The new generation of blastophagus matures in mid-July with the emergence of pollen-laden female insects (cycle 4a) (Kjellberg et al., 1987). These insects find receptive figs on female fig trees where oviposition is doomed to failure (long-stemmed flowers) (more than 95% of the insects will be sacrificed for fig pollination) (cycle 4b). The few remaining blastophagous (5%) at the end of summer (5a), enter the receptive male figs (mammae) carried by the new shoots, where they lay their eggs to complete their reproductive cycle (6a) (Kjellberg et al., 1987). For female fig trees, when the pollinating blastophages emerge from the profichis, there will then be fertilization of the long-stemmed flowers by pollen stuck to their joints during oviposition trials, leading to the development of grains in the ripe fruit in early fall (5b) (Garrone, 1998).

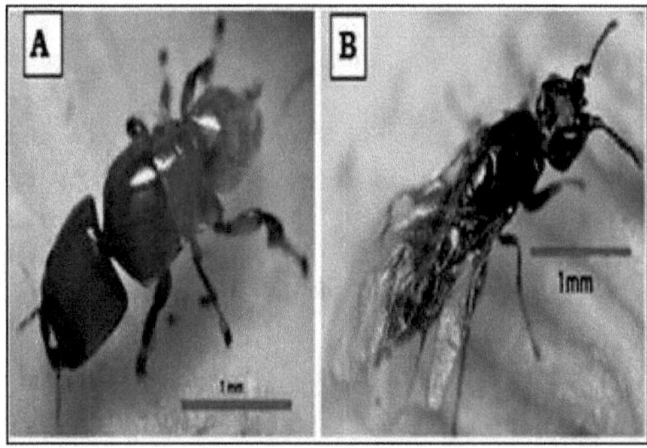

Figure 04: Female (A) and male (B) of Blastohaga psenes (Vidaud, 1997).

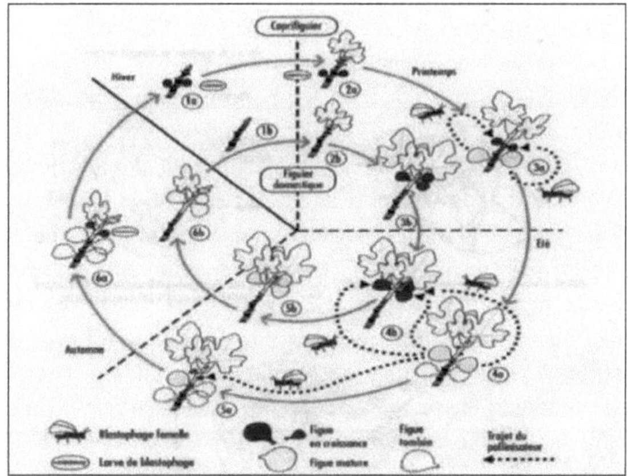

Figure 05: Simplified life cycle of the fig tree and its pollinator (Vidaud, 1997).

I.3.3. Annual vegetative and cultural cycle of fig tree

The vegetative cycle of the tree consists of three phases. It begins in February with the budburst and the formation of leafy branches and continues until May. Vegetative activity may eventually resume depending on climatic conditions and then fades by early October. The tree then begins to defoliate before entering a winter rest period of several months (Oukabli, 2003). The branching of the fig tree is done by the dormant buds of the previous year. It is of acrofuge type and builds the tree by formation of growth units in the upper part of the stem. The architecture of the tree leads to the establishment of a vigorous trunk bearing little or unbranched branches (Oukabli, 2003).

I.4. Agro-ecology of the fig tree

Although it is a temperate species commonly found in tropical and subtropical regions, the figues are now widely cultivated as interest in the plant, particularly its fruits, has reached non-native countries (Mat Desa et al.' 2019) ' but it is particularly well adapted around the Mediterranean where winters are cool and summers are hot and dry (Vidaud, 1997).Lafigue (Ficus carica L.) is a widespread species commonly cultivated, especially in hot and dry climates. The ideal condition for intensive fig cultivation is a semi-arid climate with irrigation. The fig tree is undemanding, so it adapts to a wide range of soils, from sandy to clay soils, but it prefers silty-clay soils. It tolerates pH values of 6 to 7.7 but is sensitive to high sodium and boron concentrations (Skiredj et al., 2003).Farmers must give particular importance to fig trees producers, the needs of these in terms of fertilization vary according to age (Jeddi, 2009), as an example the plants of one year should receive 9 Kg of manure well decomposed and 35 Kg of nitrogen in the form of urea, According to Vidaud, (1997), nitrogen has an action on the vegetative development and productivity of this tree, while phosphorus has an action on the quality of fruits, their coloring and ripening, potassium has an action on the yield. Pruning in spring, at the rise of sap, allows the control of the size of the tree and increases production (Lim, 2012).Despite their tolerance to drought, copious watering is necessary to improve production in quantity and quality. According to Oukabli, (2003), their real annual needs are about 600 mm, especially in spring and early summer. The average optimal temperature for growth is 18 to 20°C, but they require a temperatureAccording to Stover et al. (2007), the fig tree can tolerate freezing winter periods; it can resist to -10°C.The fig tree is multiplied mainly by cuttings which are easily rooted (Skiredj et al., 2003). The most favorable period is the beginning of March (Goby, 2006). The distance between plantations varies according to the richness of the soil, the annual rainfall and the possibilities of irrigation (Oukabli, 2003). The cuttings are spaced 20 to 30 cm apart on the line and 60 cm between lines, the plantations are made in square or along the contour lines with a distance of 4 to 6 m in all directions (Skiredj et al., 2003). The fruiting starts from the 3rd year but the maximum yield (5 tons/ha in dry land to more than 20 tons/ha in irrigated culture) is reached after 6 years (Oukabli, 2003).In the northern Mediterranean region, fig trees produce one or two crops per year depending on the cultivar. The first crop is produced from flowers that were initiated the previous year, and the fruit ripens in early summer. The second (main) crop is produced from flowers that grow on the current season's shoots, and the fruits ripen in late summer (Veberic et al., 2008).

I.5. Geographical distribution of the fig tree
I.5.1. In the world

The hardiness of cultivation, the adaptability to various environments and the easy multiplication have led to the dispersion of the fig tree in several regions of the world. Fig (Ficus carica) is one of the particular species of Ficus that spreads wild (Patil and Patil, 2011), in Southwest Asia and the Eastern Mediterranean region, from Turkey in the east to Spain and Portugal in the west; it is also grown commercially in parts of the United States and Chile and, to a small extent, in Arabia, Persia, India, China, and Japan (Chawla et al., 2012).

I.5.2. In Algeria

The fig tree is one of the three main fruit productions of Algeria: Olive, Fig and Citrus. In Algeria, the cultivation of the fig tree is ancestral, this fruit species is adapted to almost all the Algerian bioclimatic stages (Technical Institute of Fruit Growing and Vine; ITAF). The most suitable situations in Algeria, from the point of view of altitude are included between 300 and 800 m, according to the regions and the exposure, however the fig tree thrives from the littoral to 1200 m of altitude (Chaker, 1997). The majority of plantations, about 60%, are concentrated in the wilayas of Bejaia, Sétif, Constantine, Tizi-Ouzou and Bouira (Benettayeb '2018) (Table 02). The majority of the production is provided by the mountain regions of Kabylia (Bejaia, Tizi-Ouzou and Sétif) which hold respectively: 34%, 23% and 13% of the total number of trees, the latter grow the most attractive varieties such as 'Tameriout', 'Taranimt' and 'Béjaoui' (ITAF).

Table 02: Type of fig trees present in Algeria (Feliachi, 2006)

Type of fig trees	Varieties
Caprifiguiers	'Amellal', 'TitN'Tsekourt' ,'Abetroune' 'AdrasViolet', 'Azaim' ,'Medloub '
Smyrna	Alekake ', 'Amesas' ,'Tabelout ', 'Tadefouit' ''Tameriout ', 'Taranimt' 'Abougandjour ' 'Adjaffar' , 'Averane' , Avouzegar','Azendjer'
Common	'Abakor' ,'Azaich ', 'Verdale white ' 'Kadota ', 'Chetoui'

I.6. Problems related to the fig sector Algerian

The Algerian fig orchard with nearly 5 million trees is still maintained among the main fruit species of the country and constitutes more than 10% of the national arboricultural heritage. (INRAA, 2006). According to the latest FAO statistics, Algeria has experienced several fluctuations in the harvested area of figs (Figure 06). In 1988, it represented more than 50% of the rustic species other than the olive tree. This regression of areas and yields is due to several factors (INRAA, 2006) namely:

▶ The non valorization of the productions of the fig tree which makes this fruit species marginalized;

▶ The frequency of fires that ravage hundreds of hectares each year;

▶ Most of the fig orchard is aged and not very reproductive;

▶ Lack of means, materials, and non mastery of cultivation techniques;

▶ The preference of farmers for crops that provide high yields and much higher returns;

▶ The change in the eating habits of the local population has plunged the cultivation of the fig tree into various problems (ITAF);

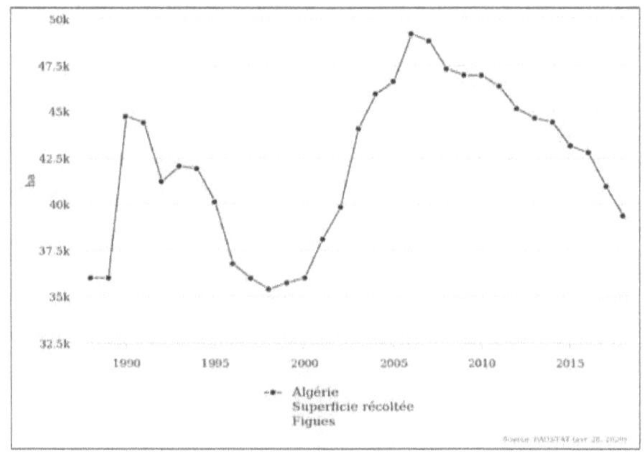

Figure 06: Evolution of the harvested area of figs in Algeria (1988-2018) according to FAO stat, (2018).

CHAPTER II
THE FIG

II.1. Description and morphology

The fig is not a true fruit (Haesslein and Oreiller, 2008), the achenes are (Chawla et al., 2012). It is found solitary and usually sessile on the branch from 5 to 8 centimeters high (Starr et al., 2003), and weight that varies between 30 to 65 grams (Ouaouich and Chimi, 2005).It is a pyramid-shaped fruit, sometimes rounded (Starr et al., 2003), spherical or ovoid with a nipple on which the tail is attached to the tree (Haesslein and Oreiller, 2008).The fig consists of a pigmented outer skin (pure green, brown, purple, or black) (Chawla et al.,2012), an ostiole (eye or operculum), and a stalk (Haesslein and Pillow, 2008). The inner side of the skin is white and contains numerous achenes (numbering 30 to 1600 per fruit) attached to the gelatinous flesh (Figure 07) (Joseph and Justin, 2011).

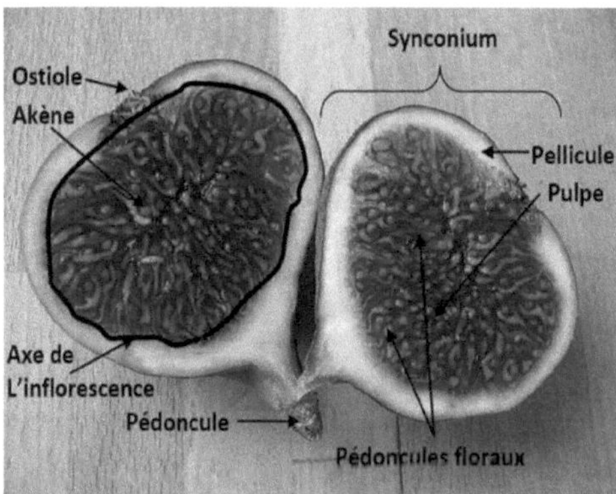

Figure 07: Morphological characteristics of the fig (Haesslein and Oreiller, 2008).

II.2. Classification of figs

The classification criteria (descriptors) were well defined by the International Plant Genetic Resources Institute (IPGRI) in 2003, based on the work of Aksoy, (1994). According to IPGRI and the International Center for Advanced Mediterranean Agronomic Studies (CIHEAM), at least 31 descriptors are considered highly discriminating (leaf and fruit shape, pollination requirement, weight, color,(e.g., height and diameter of the fruit, width of the ostiole, ease of peeling, presence of cracks in the skin, etc.).Nevertheless, these characteristics are often influenced by environmental conditions (climate and soil composition) (Chessa and Nieddu, 2005) and the antioxidant activity of some varieties of dried figs.

15

Depending on the color of the skin, the most common figs are :

• The black fig is sweet and rather dry.

• The green fig is juicy and has a thin skin.

• The purple fig is the sweetest, the juiciest, the most fragile and the rarest.

• Black- and purple-skinned varieties are eaten fresh, while green-skinned varieties are most often dried (Haesslein and Oreiller, 2008).

According to their shape, figs are divided into three groups (Bauwens, 2008):

• The first one is simply round figs, often flattened at the base and sometimes also at the top.
• A whole series of figs have a triangular or cone-shaped profile reminiscent of a pear.
• The last group consists of oblong, irregular or asymmetrical fruits.

II.3. Maturation and harvest

The ripening process of the fig is classified as climacteric, showing an elevation in the respiratory rate and ethylene production at the beginning of the ripening phase (Marei and Crane, 1971).The development of the female fruit of the fig is characterized by three phases. The first one is characterized by a rapid growth of size. During the second phase, the fruit remains almost with the same size, color and firmness. The third phase is considered the ripening phase, where the fruit develops, its color changes, the texture of the pulp softens (Marei and Crane, 1971) and its skin begins to crack.In general, fresh figs must reach a certain stage of ripeness before being harvested, as they remain immature if picked early. An unripe fig is alsoless rich in sugars and has not yet developed its organoleptic properties. On the other hand, a late harvest leads to great difficulties in handling (harvesting, storage, transportation) (Benettayeb, 2018).Fresh figs must be harvested manually and carefully in cool weather, early in the morning or late in the day. They are fragile, quickly perishable at room temperature and without the possibility of keeping them beyond two or three days in the refrigerator, their quick disposal on a local market is recommended (Jeddi, 2009).For drying, the figs are picked when fully ripe, even excessively, when they are wrinkled, semi-dry. Harvested in dry weather, after the disappearance of the morning dew, each variety is picked separately and according to its aptitude for drying (Mauri, 1952).

II.4. Production of figs

The world production of figs reached 1,135,316 tons in 2018, of which more than 90% comes from the Mediterranean basin and the Middle East (Figure 08). This production is largely based on the dry fig, which is more resistant and better storable (FAO stat, 2018). The three largest fresh fig producing countries are Turkey which alone provides about a quarter of the world's fig production with more than (306,499 t), followed by Egypt (189,339 t) and Morocco (128,380 t), (FAO stat, 2018). In the same year, Algeria holds the fourth largest world production. It represents about 10% with more than 109,214 t (Figure 09).According to FAO stat, (2018) the world harvested area of fig tree was estimated at 301,062 ha of which 61,498 return to Morocco and 51,389 ha to Turkey. In Algeria, fig plantations cover an overall area of 39 356 ha or, nearly 15% of the national tree heritage (262 000 ha), this area is occupied by more than 4.5 million trees.

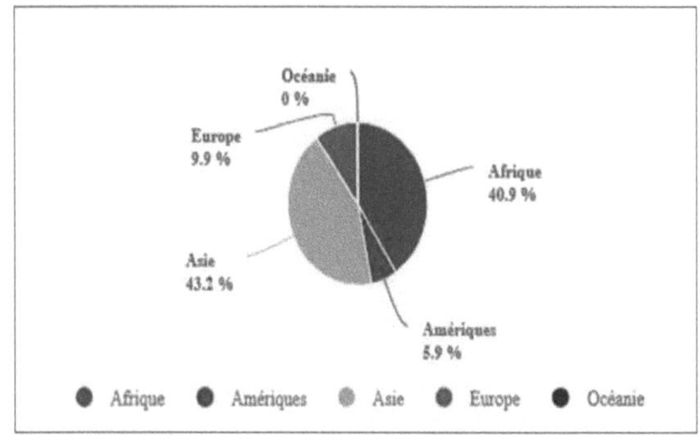

Figure 08: Share of fig production by region (FAO stat, 2018).

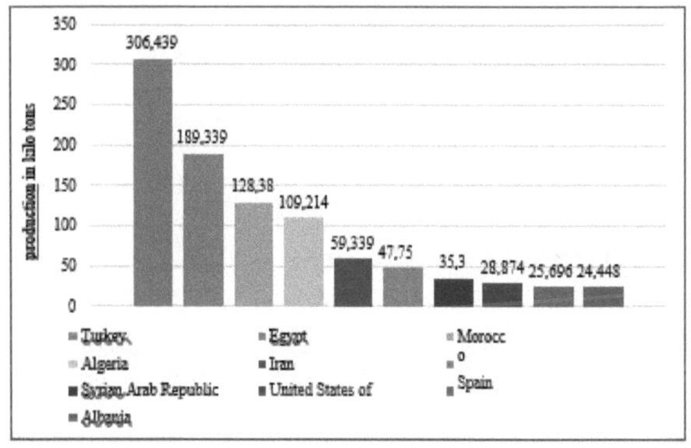

Figure 09: Fig production; top 10 producers (FAO stat, 2018).

II.5.Composition and nutritional value of the fig

The fig (fresh or dried), has a great importance in nutrition. This ancient fruit has accompanied, fed and given many pleasures to our ancestors (El Khaloui, 2010), and to justify its nutritional value there is an Arabic saying: "The day of fig no paste" (Favier et al., 1993). The fig plays a balancing role in the diet, thanks to its high content of assimilable carbohydrates that exceeds 53%, its low fat intake and the absence of cholesterol (El Khaloui, 2010). It is not only a fibrous fruit (Guvenc et al., 2009), but also an important source of vitamins, amino acids and antioxidants (Crisosto et al., 2010), minerals and trace elements, with fairly high levels of calcium, phosphorus, potassium and iron (Infanger, 2004) (Table

03).In its fresh state, the fig contains on average 80% water 13% sugar. After drying, sugars exceed 55%, it is therefore very energetic (El Khaloui, 2010).

Table 03: Nutrient content of fresh and dried figs (Arvaniti et al., 2019).

Dietary component	Value/100g fresh	Value/100g dry
Water (g)	79.11	30.05
Calories (Kcal)	74.0	249.0
Protein (g)	0.75	3.30
Total fat (g)	0.30	0.93
Saturated fat (g)	0.06	0.93
Fiber (g)	2.9	9.8
Sugars (g)	16.26	47.92
Cholesterol (mg)	0.0	0.0
Calcium (mg)	35.0	162
Iron (mg)	0.37	2.03
Magnesium (mg)	17.0	68.0
Phosphorus (mg)	14.0	67.0
Potassium (mg)	232.0	680
Sodium (mg)	1.0	10.0
Zinc (mg)	0.15	0.55
Vitamin A (IU)	142.0	10
Riboflavin (mg)	0.050	0.082

II.5.1. Carbohydrates

After water, carbohydrates are the most abundant constituents in fruits, representing between 50 and 80% of the dry weight. Carbohydrates are energy reserves and cell wall building units. Products of photosynthesis, simple carbohydrates or "sugars" are also important factors of sensory quality (Favier et al., 1993).According to Lim, (2012), a 100 g serving of dried figs provides 63.87 g of carbohydrates including 47.92 g of sugars (24.79 g glucose, 22.93 g fructose, 5.07 g starch, 0.13 g galactose, and 0.07 g sucrose). The difference in sugar content depends on the crop, maturity, storage conditions, but also can change from year to year (Gozlekci, 2011).

II.5.2. Protein

According to the table (03), the dry figs are more concentrated in proteins than the fresh figs. According to the composition advanced by Lim, (2012), the contents of "acidic" amino acids are higher than other amino acids contained in fig, whether fresh or dried.

Table 04: Amino acid composition of fresh and dried figs (Lim, 2012).

Amino acid	Fresh fig (mg/100 g)	Fresh fig (mg/100 g)
Aspartic acid	176	645
Glutamic acid	72	295
Alanine	45	134
Arginine	17	77
Cystine	12	36
Glycine	25	108
Histidine	11	37
Isoleucine	23	89
Leucine	33	128
Lysine	30	88
Methionine	6	34
Phenylalanine	18	76
Proline	49	610
Serine	37	128
Threonine	24	85
Tryptophan	6	20
Tyrosine	32	41
Valine	28	122
Total	0.64 g/100 g	2.75 g/100 g

II.5.3. Material fat

The fig contains a low amount of lipids, about 1.9% (Kolesnik et al., 1987; El Khaloui, 2010). Despite their low content, lipids have a fundamental influence on storage life, organoleptic properties and nutritional and biological value. The lipids of Ficus carica are characterized by a high rate of establishment (>68%) of monovalent fatty acids, the majority of which are polyunsaturated and which in some cases may explain the responsibility of oxidative deterioration of the fig and its derivatives (Kolesnik et al., 1987)

Neutral lipids represent the largest fraction of total lipids, their predominant compound is tri-acylglycerol with a rate of 50%, sterol esters, acid esters Fatty acids and free sterols are also present in considerable quantities. Phospholipids represent only a small fraction (Kolesnik et al., 1987). Figs contain high levels of non-essential fatty acids such as palmitic acid (16:0), oleic acid (18:1 ω 9), stearic acid (18:0) which are present in the flesh and essential fatty acids such as linolenic acid (18:3 ω3) and linoleic acid (18:2 ω6), the latter is particularly present in the bark (Guvenc et al., 2009).

II.5.4. Vitamins

The level of the vitamin varies between different parts of the fig, according to Guvenc et al, (2009), the peel and the white part of the fruit which contains a high level of γ-Tocopherol. While δ-Tocopherol is found in all parts of the fruit in different amounts, it is slightly high in the skin.Vitamins D2, D3 and α- Tocopherol acetate are found in all parts but the white part contains the highest level of vitamin D2, although the highest amount of vitamin D3 and α-Tocopherol acetate is found in the flesh (Guvenc et al., 2009). Other types of vitamins are represented in the table below.

Table 05: Composition and nutritional value of 100 g of dried figs (Guvenc et al., 2009).

Nutritional value		100 g of dried figs
Vitamins	Provitamin A	80 IU
	Thiamine (vit B1)	0.10mg
	Riboflavin (vitamin B2)	0.10mg
	Nicotinic acid (vitamin B7)	0.7mg
	Ascorbic acid	0

II.5.5. Fibers

Fresh and dried figs contain approximately 30% soluble fiber and 70% insoluble fiber (Solomon et al., 2006). Insoluble fiber promotes regular stool elimination by increasing stool volume and changing stool consistency (Haesslein and Pillow, 2008).

II.5.6. Acids

Figs contain citric acid as the majority organic acid, malic acid in appreciable amounts and acetic acid in trace amounts (Belitz et al., 2009). According to Trad et al, (2013), citric acid represents 0.35 g / 100 g fresh weight which is almost three times higher than malic acid (0.13 g /100 g fresh weight).The study conducted by Pande and Akoh, (2010) revealed that fresh fig contains in addition to the organic acids mentioned above oxalic acid (17.9 mg/100 g), ascorbic acid (14.2 mg/100 g) and succinic acid (10.2 mg/100 g), followed by traces of phytic acid (Favier et al., 1993).

II.6.Therapeutic properties of the fig

The importance of F. carica as an alternative to cure certain diseases has been recognized throughout the centuries (European Pharmacopoeia, 2007), beyond these therapeutic properties it has nutritional and cosmetic properties.That in testimony of these innumerable virtues our Prophet Mohamed (QSSL) reserved several hadiths to him, among which:

"If I were to hope for a fruit brought from heaven, it would certainly be the fig.

The main function of a fig tree is to produce delicious fruit, nutritious in a fresh or dry state, or as part of a savory or sweet preparation.Consumed raw, fig has laxative, diuretic, anti-inflammatory, cholesterol-lowering, anti-diabetic, anti-cancer and immuno-modulatory effect (Chawla et al., 2012). It also treats anemia and liver disorders, cures irritating cough and bronchitis (Kahrizi et al., 2012).Dry fig combined with acetic acid is used to treat swellings and tumors, in paste it treats burns and eczema, relieves hemorrhoids and abdominal cramps (Lansky and Paavilainen, 2011). Fig decoction is suitable for inflammatory diseases and is used to soothe coughs and stubborn colds (Clement, 1979).This fruit contains the highest level of polyphenols, flavonoids and anthocyanins (Perez et al., 2003; Solomon et al., 2006). These valuable antioxidants can protectlipoproteins against oxidation and produce a significant increase in antioxidant capacity in plasma (Vinson et al., 2005).According to Vinson, (1999), in addition to polyphenols, figs include other compounds with anti-carcinogenic activity, specifically coumarins and benzaldehydes.Fig is also known for its moderate laxative properties and can be used for the symptomatic treatment of constipation (Yancheva et al., 2005).

II.7.Phenolic profile of the fig and its power antioxidant

II.7.1. Compounds phenolics

Phenolic compounds are common secondary metabolites of plants, which not only have physiological functions in plants, but also produce positive effects for human health, as they can act as antioxidants. Phenolic compounds can serve this purpose by reducing or donating hydrogen to another compound, scavenging free radicals, and neutralizing singlet oxygen (Caliskan, 2015). The antioxidant activity of phenolic compounds is also based on chelation of pro-oxidant metal ions and inhibition of certain enzymes (Siricha et al., 2010).The polyphenols present in their structures at least one aromatic ring with 6 carbon atoms itself carrying a variable number of hydroxyl function (Ribéreau-Gayon, 1968).More than 8000 phenolic structures are known, ranging from simple phenolic molecules of low molecular weight such as phenolic acids to highly polymerized compounds like tannins (Martin and Andriantsitohaina, 2002).The phenolic fraction of the fig is defined qualitatively and quantitatively by variety, class (black, white), fruit part (pulp or skin), fruit condition (fresh or dry) (Del Caro and Piga, 2008), harvest season (June or September), origin and irrigation (Veberic et al., 2008).Piga et al, (2004) detected phenolic compounds in the skin and pulp of figs and found that the black fig cultivar had the highest content and that most of the polyphenols were concentrated in the skin.Polyphenols are very diverse compounds and can be classified into many classes and subclasses: phenolic acids, flavonoids, anthocyanins, tannins etc. (Curtay and Robin, 2000; Siricha et al., 2010).

II.7.1.1. Acids phenolics

Phenolic acids belong to two classes: hydroxy-benzoic acids and hydroxy-cinnamic acids (Veberic et al., 2008). They are among the most predominant phenolic compounds in fig and are concentrated mainly in the skin (Caliskan and Polat, 2011).According to Arvaniti et al, (2019), gallic acid, chlorogenic acid, are the most predominant phenolic acids in dry and fresh fig varieties, Similarly Veberic et al, (2008) detected in addition to these compounds traces of syringic acid.

II.7.1.2. Flavonoids

Flavonoids are the main constituents of polyphenols; more than 5000 compounds have been identified. They are widely present in almost all plants, especially in fruits and vegetables (Ribéreau-Gayon, 1968).Flavonoids are di-phenyl-propanes (C6-C3-C6) that constitute the pigments responsible for the yellow, orange and red colorations of various plant organs (Bruneton, 1987). They are classified by their chemical structure in several categories: Flavones, flavonols, flavanols, isoflavones, anthocyanidins (Zoughlache, 2009).Flavonoids behave as antioxidants in a variety of ways, including direct scavenging of oxygen species, chelation of transition metals involved in the formation of process radicals, and prevention of the peroxidation process by reducing alkoxyl and peroxyl radicals (Trifunschi et al., 2015), through inhibition of lipo-oxygenase and/or cyclooxygenase (Cheikh-Traoré, 2006).According to Veberic et al, (2008), the flavonoids present in fig are rutin, catechin, epi-catechin, etc. According to Ouchemoukh et al, (2012),dried fig is the richest fruit in flavonoids (105.6mg EQ/100g DM) compared to other dried fruits.

II.7.1.3. Flavonols

Flavonols (3-hydroxyflavones) are widely distributed, they are characterized by the presence of a carbonyl in position 4 and a hydroxyl group in position 3 (Guignard, 1996).Flavonols are present in glycosylated form (glucose, rhamnose, xylose, glucuronic acid...), they accumulate in leaves and fruits (Robards and Antolovich, 1997; Marfak, 2003).In the study conducted by Slatnar et al, (2011), 4 compounds of the flavonol group were identified in fresh and dry figs namely: kaempferol-3-O-glucoside, rutin and quercetin-3-O-glucoside, luteolin-8-C-glucoside, this last compound is identified only in dry fig.

II.7.1.4. Anthocyanins

Anthocyanins are substances belonging to the flavonoid class (Albitar, 2010). They are water-soluble pigments that participate in the coloring of some parts of plants (flowers, fruits, leaves) in blue, red, purple, pink and orange (Wang et al., 1997; Giusti and Wrolstad, 2001). They are composed of two or three parts: the aglycone base (anthocyanidin), sugars, and often acyl groups (Albitar, 2010).

Anthocyanins are very unstable molecules due to the absence of an electron in their structure, so their stability depends on pH and temperature (Wrolstad et al., 2005). The presence of two hydroxyls makes them very sensitive to oxidation, which is the reason for their antioxidant property (Castanaeda-Ovando et al., 2009).The overall anthocyanin content is considered a mark of differentiation for figs, which has a diversity of colors, ranging from dark purple to green (Solomon et al., 2006).Ouchemoukh et al, (2012) showed in their research that dried fig is the richest fruit in anthocyanins compared to other dried fruits such as plums (2 mg/100g DM) and grapes.The species Ficus carica is characterized by the presence of the majority of 2-monoglucosides of anthocyanidins, in particular cyanidin 3-glucoside and keracyanin (Del Caro and Piga, 2008), in addition to these compounds, Duenas et al., (2008) have detected: cyanidin-3,5-diglucoside and pelargonidin 3-rutinoside.

II.7.1.5. Tannins

Tannins are by definition polar phenolic compounds of plant origin (Berthod et al., 1999) with molecular weights between (500-3000 Daltons), they exist in almost every part of the plant (bark, wood, leaves, fruits and roots) (Cowan, 1999; Zimmer and Cordesse, 1996).Previous studies have been conducted on polyphenolic extracts of F. carica and reported many bioactive compounds, namely good tannin richness (Al-Snafi, 2017; Sharma et al., 2017).The most determining characteristic of tannins is their capacity to form complexes (by precipitation) with natural polymers such as proteins, polysaccharides (pectin, cellulose, hemicellulose...), alkaloids, nucleic acids and minerals (Frutos et al., 2004).Tannins can act as antioxidants. The antiradical capacity of procyanidin dimers and trimers is increased with galloylation and to a lesser extent with chain length, it is also influenced by the position of the galloyl substituents (Cheynier, 2005; Gramza and Kolczak, 2005).Similarly Al-Maliki, (2012) reported in his previous studies that fig has antimicrobial activity from certain active tannic compounds such as ellagic acid, chebulic acid and gallotanin.

A. Condensed tannins

Condensed tannins (CTs), or pro-anthocyanidols (Muller Harvey and Mc Allan, 1992; Bruneton, 1999) are flavan polymers consisting of flavan-3-ols (catechin and epi-catechin) and flavan 3,4 diols most often linked together by C4-C8 bonds, the most common class are the procyanidins. Debib et al, (2013) reported in their study that fig had a condensed tannin content between 10 and 194 mg EAG / 100 g.Like all phenolic compounds, pro-anthocyanidins also have an antioxidant power thanks to their redox potential and free radical scavenger activity (Kelm et al., 2005).

B. Hydrolysable tannins

Hydrolyzable tannins are polyesters of a sugar (mostly glucose) and a variable number of phenolic acid molecules which are considered as non-flavonoids (ellagitanine, gallotanins) (Delluc, 2004).By hydrolysis (acid, alkaline or enzymatic), the phenolic acids released are gallic acid or ellagic acid, from gall tannins (gallotanins) and ellagic tannins (ellagitanins), respectively (Zimmer and Cordesse, 1996).

II.8. Carotenoids

Carotenoids are natural pigments that belong to the tetraterpene family and are represented by more than 600 known natural structural variants, they contribute to the yellow, orange or red coloring of fruits and vegetables (Tapiero et al., 2004).These lipophilic pigments contain a cyclic structure at each end, which gives them the ability to receive a free radical without losing their stability; they therefore have a notable antioxidant effect (Dionne, 2002). There are two main classes of carotenoids:

o Carotenes, non-polar, they have only hydrocarbon chains (α- carotene, β-carotene); they confer the orange and red colors.

o Xanthophylls, are the most polar because of the presence of oxygen in their structures (lutein, β-cryptoxanthin, zeaxanthin, astaxanthin, ...), they are responsible for the yellow color (Dionne, 2002).

Ouchemoukh et al, (2012) noted in their work that Ficus carica is one of the sources of carotenoids with a content of about 11 mg EβC/100g DM.The carotenoid profile of fig studied by Kakhniashvili et al, (1987) showed that lutein and α-carotene are the most dominant carotenoids, in addition to the presence of violaxanthin, neoxanthin, rubixanthin and kryptoxanthin in lower content. Belitz et al, (2009) reported the presence of other compounds which are; phytoene, phytofluene and luteoxanthin.

CHAPTER III

DRYING OF THE FIG

III.1. Drying of figs

Fruit drying is one method of preserving food over an extended period of time (Okos et al., 1992). This process remains a common practice, especially for local products (Cantin et al., 2011). However, a poor mastery of this technique leads to the loss of nutritional and therapeutic values of the food (Piga et al., 2004).Fresh figues are highly perishable in the ambient state. At low temperature and high humidity (4.44 - 6.11°C, 75% relative humidity), figues can remain in good condition for 8 days, however, when stored under ambient conditions, they can only last 1-2 days (Mat Desa et al., 2019). Their post-harvest life varies depending on the variety, temperature, and degree of ripeness at the time of harvest (Doymaz, 2005).

III.1.1. Objective of the drying

Many fruit varieties are seasonal and their availability for extended consumption has been made possible by drying (Mat Desa et al., 2019). Drying is a process of removing moisture through the action of heat. It serves as preservation by inhibiting the activity of water in fresh products (Mat Desa et al., 2019). On the one hand, it reduces the weight and volume of the fig, as well as the packaging and cost of delivery, microbial activity and chemical changes during storage (Martinez-Garcia et al., 2013), and on the other hand it increases the concentration of total polyphenols and consequently the antioxidant activity but, decreases the concentration of anthocyanins (Martinez-Garcia et al.,2013).Drying also makes it possible to valorize food products into stable dried products by absorbing the overproduction (Mamouni, 2002).

III.1.2. Fig dry

Only pollinated figs are suitable for drying, hence the interest of caprification (Mamouni, 2002; Oukabli, 2003), while flowering figs and autumn figs are varieties intended only for fresh consumption (Mamouni, 2002).The most valued criteria for figues intended for drying are: high sugar and soluble solids content, low acidity and soft skin to produce Dried figues with light color, soft texture and sugar-rich fractions (Mat Desa and al., 2019).Black and purple-skinned varieties are consumed fresh, while green-skinned varieties are most often dried. After drying, the latter become white (Mamouni, 2002; Oukabli, 2002), while the colored figs retain their color (Mamouni, 2002).The standard states that dried figae produced for direct consumption should contain a moisture content of 26% for untreated figae and 26-40% for high-moisture treated figae (Mat Desa et al., 2019).Dried fruits are durable products due to their low water activity, the main problems in the dried fig trade are storage pests and mycotoxins. To ensure consumer safety, dried products must be free of chemical contamination as well as bacterial and fungal infestation (Mat Desa et al., 2019). Dried figs are more spoiled by Aspergillus sp, Alternaria sp, Botrytis cinerea, Fusarium sp, Saccharomyces, Pichia, and Bacillus (Cantin et al., 2011). One of the common types of mycotoxins detected is aflatoxin that develops from A. flavus infection (Mat Desa et al., 2019).The final quality of the dried fig depends on the drying parameters such as temperature, speed and relative humidity of the drying air, as well as the duration of drying and also their conditioning, duration and storage conditions. More importantly, the final product quality in dried figues depends on the maturity stage of the fruit, as sugar accumulation and ripening occurs during the last three days (Mat Desa et al., 2019).The color and firmness of the fruit are the criteria generally used to determine the optimal harvest date. Figs for drying should be picked very ripe, they should be harvested in dry weather and each variety should be picked separately according to its aptitude for drying. The perfectly ripe fig wilts, its bearing is no longer erect, the skin is slightly cracked; the stalk, initially turgid and milky white, becomes dry and translucent, so it is easily detached with its stalk, unlike an underripe fig (Ouaouich and Chimi, 2005).

III.2. Methods of drying

The figs are harvested with a very high moisture content, which is conducive to various degradations, making the products very perishable. Consequently, for lack of means of conservation, the losses can be very high (Ouaouich and Chimi, 2005).To extend the shelf life of figs and their availability during all seasons, farmers use traditional drying methods, while at the industrial level they use faster artificial drying processes to ensure a large production for trade.

III.2.1. Drying traditional

Sun drying is a traditional preservation method (Sen et al., 2010; Faleh et al., 2012) used to obtain dried figs without financial expense and with simple equipment. This method provides figs with good taste and consistency (Faleh et al., 2012).It is a technique based on natural convection through the circulation of hot ambient air. Convective drying consists of the removal of water from the surface of the fruit by the hot air, which creates a gradient inside the fruit encouraging the water to move by diffusion from the interior to the surface (Carranza-Concha et al., 2012).In areas where climatic conditions are adequate (Sen et al., 2010), figs are spread in monolayer under the sun (El Khaloui, 2010), in the open air (Oukabli, 2002), on the dirt floor, on terraces (Jeddi, 2009), on the roofs of buildings (El Khaloui, 2010), or on mats or reed racks (Gamero, 2002).For a regular drying, the fruits must be turned over every day. To avoid deterioration, the trays are put under cover in the evening. Drying lasts 3 to 6 days, depending on the temperature of the season (El Khaloui, 2010). The figs are considered dry when they acquire an elasticity to the touch and do not leak syrup under the effect of pressure between the thumb and index finger (El Khaloui, 2010).However, sun-drying exposes the products to dust, flies and dirt, and to numerous and varied contaminations. In addition, this method, which costs practically nothing, does not allow any control over the drying parameters and extends the drying period. of drying. As a result, the quality of the product is very poor in terms of hygiene and nutrition, so it is not recommended for economic and especially public health reasons (Belaid, 2015).

III.2.2. Drying artificial

Today, a significant amount of fig production is dried, based on modern techniques, by mechanical drying systems with hot air (Babalis et al., 2006).The raw materials to be dried are always subjected to a preliminary preparation (cleaning, sorting, grading, bleaching, fumigation, etc.) for further processing. These preparation operations vary according to the nature of the raw material and the product to be obtained (Ouaouich and Chimi, 2005). The main ones are mentioned in the diagram below (Figure 10).

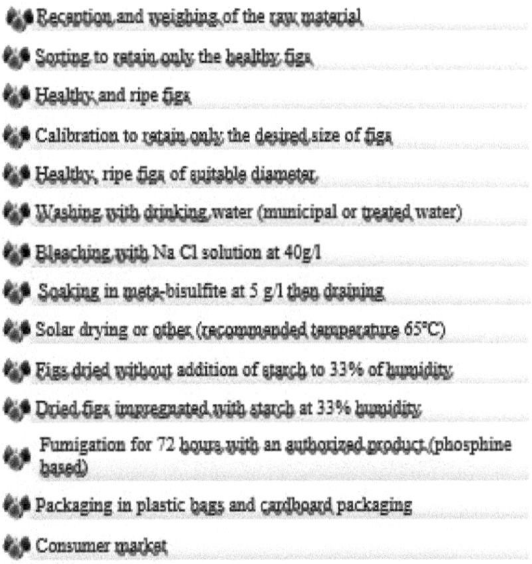

Figure 10: Manufacturing flow chart of dried figs (Belaid, 2015).

III.2.2.1. Reception

At the reception, the figs must be weighed. In order to ensure the quality of the fruits received we must proceed to the identification of the varieties of figs. The percentage of foreign material and damaged and contaminated fruits gives an idea about the condition of the raw material (Belaid, 2015).

III.2.2.2. Sorting and calibration

In the drying unit, the sorting is done manually on inspection belts. A manual sorting allows to eliminate all the fruits unfit for consumption, not very ripe, damaged, too big or too small. The grading consists in obtaining fruits of the same size (same volume, same density) to ensure a uniform behavior during the drying process and to couple to the quality of the finished product a homogeneity for its presentation. From a batch of the same variety, the fruits are classified into different sizes on the basis of equatorial diameter or volume (Belaid, 2015).This sizing can be done by hand or using an industrial size, but for figs there is no regulated industrial size. Non-conforming fruits are preferably destined for the manufacture of jam or other candied products (Belaid, 2015).

III.2.2.3. Cleaning and washing

The purpose of this step is to eliminate contaminated fruits and any dirt and foreign matter in order to obtain fresh and clean products. In particular, it allows to reduce the drying time and to exploit in better conditions the capacity of the dryer (Belaid, 2015).

III.2.2.4. Bleaching

The preliminary blanching is intended to clean the skin by removing the dust that covers it, the traces of latex that make the fruit sticky and the skin more permeable, which facilitates drying. It is obtained by sprinkling for 20 to 30 seconds with soda water (1% soda) heated to 80°C on the stainless steel mesh belt. This sprinkling is followed by a rinsing by sprinkling hot water slightly acidulated with citric acid to eliminate any trace of soda. The figs should be, as much as possible positioned "tail up" on the mat to prevent the soda from penetrating the fruit through the ostiole (Belaid, 2015).

III.2.2.5. Drying itself

The drying is done in mechanical dryers in a closed enclosure to control the drying parameters, optimize energy and ensure the product safety and quality standards required (Ouaouich and Chimi, 2005 ; Jeddi, 2009 ; El Khaloui, 2010).

Industrial drying is based on the use of conventional dryers (ovens) or hybrid dryers (oven and solar) for dehydration (Figure 11), the latter has several advantages of which the most important are (Ouaouich and Chimi, 2005):

■ The product is dried indirectly with ventilated air which avoids the degradation of its photon sensitive ingredients;

■ Very good final quality of the dried product;

■ The dryer uses the energy of the sun during the day and of the fuel oil at night which avoids the rehydration of the product during the night;

■ Low installation cost;

■ Ease of construction;

Figure 11: Photograph of a hybrid dryer (Ouaouich and Chimi, 2005).

Immediately after the soaking operation, the figs are spread in a single layer on racks that are superimposed in the enclosure of the dryer (El Khaloui, 2010).The hybrid dryer uses mainly indirect solar energy and a back-up source using gas or diesel which is activated at night and in cloudy weather beyond the minimum required to avoid rehydration of the products. Solar energy is collected by collectors installed on the roof and routed to and from the dryer compartments by flexible polyethylene tubes. A ventilation system powered by solar cells propels hot air into the different parts of the dryer (Ouaouich and Chimi, 2005). The drying temperature is set between 60 and 65°C (El Khaloui, 2010), but the first phase of drying should be done at temperatures below 45°C (Gamero, 2002). The racks are occasionally removed and turned over to improve the drying conditions. The operation takes about three hours, after which the figs acquire the desired golden yellow color (Figure 12). The cooled figs are fumigated with a phosphine or carbonic acid based product to avoid the development of insect larvae (Ouaouich and Chimi, 2005).

Figure 12: Photograph of dried figs (Ouaouich and Chimi, 2005).

III.2.2.6. Packaging, wrapping and storage

Deterioration of color, flavor, and texture is possible both before or during drying as well as during storage, so conditioning is required (Belaid, 2015).Dried products are sorted according to moisture (roasted and overly hydrated fruits are eliminated), size, and color and then weighed before being placed in polyethylene or polyvinyl-based food packaging, cellulosic film (cellophane), or paper and cardboard packaging. Packaging and wrapping allows (Belaid, 2015):

• Preserve the color and aroma of the fruit and keep them away from humidity to avoid any external contamination.

• To slow down as much as possible the deterioration reactions of the product, provided that a suitable packaging is used which allows to maintain the low level of water activity reached at the end of the drying process and the temperature maintained inside at about 25°C.

• Give them an attractive appearance and facilitate their handling and storage.

III.2.3. Other methods of drying

III.2.3.1. Dehydration osmotic

Osmotic dehydration is a drying technique that involves partially removing moisture from the product by placing it in a concentrated sugar solution (Naikwadi et al., 2010). The fruit absorbs some of the sugar and is able to retain more water at the end of the process, making it softer than if it had only been air-dried (Ife Fitz and Bas, 2003).Fruits dried in this manner have a porous, crisp structure and retained a high percentage of their odorous volatile compounds (Naikwadi et al.,2010).This technique was the subject of the work of Naikwadi et al, (2010), the clean figs were subjected to a steam treatment at 90°C/5 min in autoclave and then soaked in different syrups of sucrose, glucose, fructose and invert sugar at 50° Brix for 24 h. After removal from the sugar solutions, the figs were dried at 50-55°C for 18-20 h in order to bring the fruit moisture to 18-20%.The results showed that figs prepared in fructose and invert sugar syrups maintained their sensory and hygienic quality for 6 months compared to untreated dried figs.

III.2.3.2. Drying by micro waves

In recent years, microwave drying has gained popularity as an alternative drying method in the food industry due to its uniformity and selectivity (Sharifian et al., 2012).It is a fast, more consistent method and offers significant energy savings with a potential reduction in drying time of up to 50% in addition to inhibiting the surface temperature of the treated material compared to conventional hot air drying (Abou-Farrag et al., 2013). Microwaves are part of the electromagnetic spectrum, their frequencies are between 300 and 300,000 MHz with corresponding wavelengths between 1,000 and 1 mm. Microwave energy generates heat in products with a high water content. In other words, the microwaves heat the free water, causing it to be removed from the product as steam. Fig contains internal free water which allows the application of this method (Sharifian et al., 2012).

CHAPTER IV

PREVIOUS STUDIES DEALING WITH THE IMPACT OF DRYING ON THE PHYSICOCHEMICAL PROPERTIES, NUTRITIONAL QUALITY AND ANTIOXIDANT ACTIVITY OF FIGS (FICUS CARIA L.)

Considering on the one hand, the exceptional circumstances that the whole world and particularly our country are going through because of the pandemic of the Coronavirus (COVID-19) and the partial quarantine that affected our wilaya we can only do without the experimental part that was to take place in our thesis. On the other hand, since no study has been carried out on the fig of our region (Jijel), the research works already carried out in Algeria and abroad on the same theme have been passed in this part in the form of a synthesis document.It is known that fruits are truly among nature's great gifts as they provide many essential nutrients for the health and maintenance of our bodies. The fig (Ficus carica L.) is a delicious and nutritious fruit, it has been used since ancient times along with other parts of its trees for medicinal purposes. Figs can be for fresh consumption, but are also very popular as dried fruit or valued as candied (Abul-Fadl et al., 2015).Many varieties of fruits need to be processed to maintain their quality as they are only offered in season, among these fruits we distinguish figs (Manoj et al., 2018). These are also very susceptible to microbial deterioration due to their high water content, which presents an obstacle to their preservation even under cold storage conditions. Thus, reducing the moisture content to a level that allows safe storage over an extended period of time through a drying process can solve the sensitivity problem of figs (Doymaz et al.,2004; Faleh et al., 2015).Product drying is an ancient practice for food preservation that is still widely used (Kamiloglu and Capanoglu ,2015). Among the different drying techniques, sun drying, air drying, and microwave drying are the most studied methods.

IV.1. Drying at sun

Despite many disadvantages, sun drying (traditional drying) is still practiced in many places around the world such as tropical and subtropical countries. Solar energy is an important alternative energy source and is preferred toIt is abundant, non-fuel and non-polluting. In addition, it is renewable, cheap and environmentally friendly (Doymaz et al., 2004). The most important quality of figs known to be affected by high temperature drying (sun drying) for a long time includes: physico-chemical properties, nutritional value, and phytochemical composition.

IV.1.1. Influence of sun drying treatment on the characteristics physicochemical

In a recent study conducted by Chauhan et al, (2015) the physical properties of an Indian fig variety including length, width and density were estimated before and after sun drying. The first two parameters were measured using a caliper, while density was measured by the toluene displacement method, the latter is based on immersing the fig samples in a graduated cylinder containing toluene. Density is estimated as the ratio of sample mass to volume of toluene displaced.According to the results obtained, the length of fresh and dried figs was between 15.46 mm, and 14.26 mm, respectively. The width of fresh samples was equal to

18.14 mm and 17.46 mm for dried samples.Regarding density, the values obtained were 0.93 g/$cm3$ and 0.94 g/$cm3$ for fresh and dry figs, respectively. Similar results were reported by Manoj et al, (2018) where they showed that the sun drying process had a direct influence on the different physical parameters of the Indian fig samples due to the decrease in water content as the operation progresses, as well as to the variations in the volume mass and structure of the cell wall.Moisture is an important complementary parameter to know the water content and to estimate the yield after drying of fruits (Ribéreau-Gayon, 1968). In addition, its determination is essential, because it influences the structure, appearance and taste of the fruits (Cendre, 2011).Nakilcioglu and Hisil, (2013) evaluated the moisture content of about ten samples of fresh and dried figs of the variety "Sarilope" present in Turkey. The moisture content was determined by drying the samples in an oven at 103 ± 2 C° until the following was obtained stable weight (AOAC, 1997). In the fresh state, the moisture content varied between 76.44% and 82.69 %. After drying, t h e moisture content of the samples decreased to values between 16.73% and 27.89%.Manoj et al, (2018) reported in their study conducted on Indian figs that the moisture content of fresh fig was 80.2% and that of dried fig 25.86%. The results of the present study are slightly higher than those found by Ait Haddou et al, (2014) who evaluated the moisture content of seven local fig cultivars in Morocco before and after drying. They reported that the moisture values for fresh figs ranged from 54.57% to 73.40%, while for dried figs they ranged from 20.25% to 24.30%.Among the studies conducted on dried fig, Al Askari et al, (2012) examined the moisture content of 72 samples taken from the markets of rabat and Casablanca in Morocco, where they reported in their result an average moisture value of 21.7%.Overall it was concluded that all the water contained in fresh figs can not be removed by the drying process, this is due to the fact that the fruits are composed of free water and bound water, the latter remains attached to hydroxyl, carbonyl and amino groups of simple sugar molecules, polysaccharides ... etc (Al Askari et al., 2012).Ash is the amount of minerals present in a sample or substance. Ash content is one of the strategies used to determine the amount of minerals offered in a particular sample (Manoj et al., 2018).According to the AOAC method (2000), the determination of ash consists of incinerating samples in a muffle furnace at high temperature until a black color is obtained and white smoke appears.Manoj et al, (2018) evaluated the ash content in fresh and dry figs, they reported higher values in sun dried samples (4.42%) compared to fresh samples (4%). These results were consistent with those found by Chauhan et al, (2015) who reported a value of 4.4%.Similarly in another study conducted by Soni et al, (2014) on a variety of fig commonly known as "Anjir" in India, the ash content detected after drying was almost similar to the results of the cited work with a value of 4.65%.Titratable acidity and pH are important elements for the determination of the date of harvest and even the degree of maturity of the fruits (Ouaouich and Chemi, 2005; Chahidi et al., 2008).The principle of determination of total titratable acidity according to AOAC, (1990) is to perform a titration by a standardized solution of NaOH to 0.1N in the presence of a colored indicator, and the results are expressed by convention in grams of citric acid for fig. While pH is measured by a pH meter (Al Askari et al., 2012)Regarding titratable acidity, Al Askari et al., (2012) reported values ranging from 0.26 to 0.36 g citric acid /100g dry matter (DM). These results are lower than those reported by Bachir Bey et al, (2016) who reported levels ranging from 0.77g to 1, 92 g citric acid /100g DM.In the study conducted by Hoxha and Kongoli, (2016) on two varieties of Albanian dry figs named "Roshnik" and "Malakuq"؛ the pH value varied between 4.21 and 4.35. These data are lower than those obtained by Al Askari et al, (2012) where they reported values ranging from 4.9 to 5.4.

IV.1.2. Influence of sun drying treatment on the nutritional value

Carbohydrates constitute the major part of our diet and are provided mainly by fruits (Lee et al., 1970). The carbohydrate concentration of fruits is of great interest, because of their influence on organoleptic properties and is a criterion for evaluating ripeness, it also conditions the stability and preservation of fruits (Golubev et al., 1987; Jiang et al., 2013).Among the methods for carbohydrate assays is the anthrone method. The latter has been adopted in several works including those of Manoj et al, (2018) and Chauhan et al, (2015). Its principle consists first of a hydrolysis of carbohydrates into simple sugars using dilute hydrochloric acid. Then in a hot acidic medium, glucose is dehydrated to hydroxymethyl furfural, this compound forms with anthrone a green colored product with an absorption maximum at 630 nm.The results of the study conducted by Chauhan et al, (2015) reported carbohydrate contents of 16.3 g /100g for fresh figs and 65.15 g /100g for dry figs. These results are in agreement with those found by Manoj et al, (2018).In addition to the anthrone method, other authors have used the HPLC technique for carbohydrate determination, including Faleh et al, (2015) who studied the effect of sun drying on the sugar content of ten Tunisian cultivars of different colors. Carbohydrate extraction was performed by heating the samples with aqueous ethanol (80% v/v), the mixture was evaporated to dryness and the solvent was removed by vacuum evaporation. The residue obtained was dissolved and filtered and analyzed by HPLC (Miguez Bemardez et al., 2004). The results revealed that the dry fig extracts had a sugar content of about 34.064 g /100 g. Fructose and glucose were identified as the main monosaccharides in dry fig, with contents between 9 and 31 g/100 g and between 12 and 19 g /100g respectively.Faleh et al, (2015) reported that the observed differences in glucose and fructose levels could be associated with the activity level of enzymes involved in sugar metabolism, such as invertase and sucrose phosphate synthase. In another study conducted by Slatnar et al, (2011) on local varieties of the sun-dried figs known as "Belapetrovka" in Slovenia, the HPLC method was also used to determine the sugar content. The results were lower than those found by Faleh et al, (2015) with an average content of 22.9 g / 100 g. Slatnar et al, (2011) also reported that fructose and glucose were the most dominant sugars in all analyzed samples with contents of 12.1 and 10.3 g /100 g, respectively.Food proteins of plant origin occupy a large part of the human diet even if they have a lower biological value than proteins of animal origin (Lacroix, 2008).For protein determination of dried figs, Chauhan et al, (2015) and Soni et al, (2014) examined the protein content by using the Kjeldhel method. In the assay, the sample is mineralized in an acidic medium (using sulfuric acid at high temperature) where the protein nitrogen is converted to ammonia nitrogen in the presence of a salt (potassium sulfate) and a catalyst (selenium, mercury oxide II, copper sulfate II). The ammonia is then released as a salt (ammonium sulfate) by addition of a concentrated solution of NaOH in excess before being distilled by steam and trapped in a boric acid solution. The ammonium borate is titrated with a standardized solution of acid (hydrogen chloride or sulfuric acid) and an indicator.

Chauhan et al, (2015) reported that dry fig contains only 3.01% protein, while a contribution of 4.67% was determined by Soni et al, (2014). These results are lower than those obtained by Ait Haddou et al., (2014) who found values ranging from 4.17 to 7.23% for the different fig varieties analyzed.Fibers are parts of plant origin that our body does not digest, they play however an important physiological role in the intestinal transit.Very few studies have been performed to determine the fiber content in dried figs. According to the method described by AOAC 985.29 and adopted by Soni et al., (2014), the principle of fiber determination consists in a first step of digestion of the samples with an enzyme cocktail (α amylase, protease and amyloglucosidase), followed by treatment with ethanol to precipitate soluble fibers and extract proteins and glucose. The resulting precipitate is washed, dried and weighed.The results of Soni et al, (2014) showed that the dietary fiber content of dried figs is 3.68%. In addition the data from the study conducted by Lim, (2012) on Malaysian figs reported a content of 2.9 g of total fiber per 100 g of fresh figs and 9.8 g per 100 g of dry figs.Fig is known to be an excellent fruit for health due to its negligible fat and cholesterol content. These facts were confirmed by the work of Chauhan et al, (2015). For the determination of fat they adopted the Soxhlet method described by the AOAC, (2000). This is a reference method in which total fat (TFM) is extracted using ethyl ether, after evaporation of the solvent the residue obtained is weighed and the fat is determined. Inputs of 0.53% for fresh fig and 0.56% for dried fig were obtained at the end of this study. These results were supported by those obtained by Soni et al, (2014) for dry figs (0.56%).Overall it was concluded that sun drying increases the nutrient content due to the removal of water content.

IV.1.3. Influence of sun drying treatment on phenolic compounds

Phenolic compounds are present in all fruits as a diverse group of secondary metabolites (Kamiloglu and Capanoglu ,2015; Oliveira et al., 2009).Figs have good levels of nutrients and are an important source of polyphenols that contribute to its quality. In addition to antioxidant effects, phenolic compounds have a wide range of biochemical properties and may also have a beneficial effect on the prevention of the development of different diseases (Bachir Bey and Louaileche, 2015; Abul-Fadl et al., 2015).In the literature, over the last decade, scientific interest has focused on the qualitative and quantitative analysis of phenolic compounds including phenolic acids, flavonoids and anthocyanins in different varieties of fresh and dried figs.Extraction is one of the most important steps for the purification and pre-concentration of specific compounds and plays a crucial role in the isolation and qualitative analysis of phytochemicals (Arvaniti et al ., 2019).Several parameters can influence the extraction of phenolic compounds including their chemical structure, time and storage conditions as well as the presence of interferents, the size of the particles forming the sample and the extraction solvent (Naczk and Shahidi, 2004).Different organic solvents such as acetone (Bachir Bey and Louaileche, 2015), ethanol, methanol (Manoj et al., 2018; Chauhan et al.,2015; Bachir Bey et al.,2016; Faleh et al.,2012; Hoxha and Kongoli, 2016; Nakilcioglu and Hisil, 2013; Pourghayoumi et al, 2017) and their combinations with water (Hoxha et al., 2015; Faleh et al.,2015; Kamiloglu and Capanoglu, 2015) have been widely used to recover different phenolic compounds from fresh and dry fig. Moreover, aqueous solvents give the best extraction yields than absolute solvents (Spignon et al., 2007). It was also shown that among the four solvents, methanol 80%, ethanol 80%, water and acetone 80%, the acetone/water combination was the mixture that gave the highest levels of phenolic compounds (Bachir Bey et al., 2013).The most commonly used methods for the determination of total polyphenols in fig extracts are by means of colorimetric assays including the Folin- Ciocalteu method which has been applied by many researchers (Hoxha et al., 2015; Faleh et al.,2015 ; Kamiloglu and Capanoglu, 2015; Manoj et al., 2018; Chauhan et al.,2015; Bachir Bey et al.,2016; Nakilcioglu and Hisil , 2013; Pourghayoumi et al, 2017; Bachir Bey and Louaileche, 2015). This method is based on a redox reaction, where the Folin- Ciocalteu reagent is reduced, upon oxidation of phenols to a mixture of blue oxides of tungsten and molybdenum. The coloration produced, with a maximum absorption at 750 nm, is proportional to the quantity of polyphenols present in the plant extracts (Boizot and Charpentier, 2006). Gallic acid (GA) or tannic acid (TA) are used as standards and the phenolic content is expressed as mg of gallic acid equivalents (GAE) or tannic acid equivalents (TAE) per 100 g of fresh or dry sample (Arvaniti et al., 2019).In recent years, several studies have also used more specific and selective analytical methods for the determination of phytochemicals in fig extracts (Faleh et al., 2012). The most commonly used instrumentation for the qualitative and quantitative determination of phytochemicals in fig samples is high performance liquid chromatography (HPLC) coupled with spectrophotometric detection such as ultraviolet-visible (UV-Vis) or diode array detection (DAD) and mass detection (MS). In most cases, these analytical methods are useful for the identification of these compounds from different parts of figs / whole fruits. In addition, these analytical techniques can reach very low detection limits for some phytochemicals. Liquid chromatographic separation of these phenolic compounds was mainly performed using C18 and C8 columns (Arvaniti et al., 2019).The total polyphenol content in fresh and dried fig of some varieties grown in India was measured by Manoj et al, (2018). The results of these analyses showed that the polyphenol content in fresh fig extracts was 4.58 mg EAT/ 100 g and 4.92 mg EAT/100 g after drying. In the research conducted by Slatnar et al, (2011) on a fig variety grown in Slovenia, a total phenol content of 7.49 mg EAG /100 g was reported in fresh fig fruits with a significant increase after drying (49.5 mg EAG/ 100 g). According to the reports of these studies, it had been found that the phenolic content increased after drying. This is due to the loss of moisture and also to the fact that drying is responsible for releasing the binding of phenolic compounds from the matrix throughout the degradation of cellular constituents.Contradictory results were reported in the study conducted by Bachir Bey et al, (2016) on three local fig varieties from Algeria, where they indicated a regression in the Total phenol content ranged from 107.08 - 181.06 mg/100 g DM in the fresh state to concentrations ranging from 30.81 to 40.91 mg/100 DM after drying.Overall, the variability in polyphenol content among the cultivars studied may be due to many parameters such as geographical origin, weather and post-harvest storage conditions, varieties, extraction conditions, etc...(Bachir Bey and Louaileche, 2015). Regarding the influence of color on the content of phenolic compounds of fig (Ficus carica L.), Kamiloglu and Capanoglu, (2015) reported in their study on two varieties of dried Turkish figs of yellow and purple color a content of 193 mg EAG / 100 g and 417 mg EAG / 100 g for the yellow and purple fig, respectively. Similarly in another study conducted by Bachir Bey and Louaileche, (2015) on nine varieties of Algerian dried figs including three dark and six light colored, the results indicated that the content of phenolic compounds of dark varieties was higher than that of light varieties with average values of 618.85 mg EAG/100 g and 514.72 mg EAG / 100 g, respectively.In addition other researches have focused on studying the influence of other parameters such as assay methods on phenolic content. Indeed, this was shown in the study carried out by Faleh et al, (2015) who evaluated the content of total polyphenols by the Folin-Ciocalteu method in five Tunisian fig cultivars. After drying they noted contents ranging from 121.3 to 277.2 mg/100 g. On the other hand in another study Faleh et al, (2012) determined by HPLC-DAD the phenolic profile of the same varieties of dried figs, where they mentioned lower concentrations of total phenolic compounds compared to those determined by the reagent of Folin-Ciocalteu with contents ranging from 14.22 to 78.14 mg / 100 g.The difference in results between the two methods is due to the fact that the Folin- Ciocalteu method detects not only polyphenolic compounds but also other biological compounds that have reducing power, especially sugars, aromatic amines, sulfur dioxide, ascorbic acid, organic acids and Fe (II) etc... (Faleh et al., 2015).Among the most analyzed phenolic compounds in fresh and dried figs are anthocyanins. These are pigments especially associated with fruits (Peterson and Dwyer, 1998) and they play a very important role in pollination (Hiem et al., 2002). They are also very sensitive compounds and can be destabilized by several factors including high temperature and light (Laleh et al., 2006).Several works have been carried out to evaluate the anthocyanin content in fresh and dry figs using the differential pH method (Kamiloglu and Capanoglu ,2015; Chauhan et al., 2015; Hoxha et al., 2015; Pourghayoumi et al., 2017). It is a spectrophotometric method that allows a fast and accurate measurement of total anthocyanins even in the presence of degraded polymerized pigments and other interfering compounds. It is based on the determination of the absorbance of extractive solutions diluted with buffer solutions of pH = 1 and pH = 4.5. Anthocyanins are reversibly transformed under the influence of pH. The structural change associated with the modification of the chromophores determines the different color of the anthocyanin solutions as a function of pH. The colored form (oxonium) predominates at pH 1 and the colorless form (hemiacetal) at pH 4.5. Absorbance is measured at 520 and 700 nm and results are expressed as mg cyanidin-3-glucoside (C3G) per 100 g fresh or dry sample (Ali Rachedi et al., 2018) .Quantitative estimation showed a decrease in anthocyanin content after drying according to the results obtained by Kamiloglu and Capanoglu, (2015), where they reported a content of 4.6 mg EC3G/100g and 0.1 mg EC3G/100g for fresh and dried fig, respectively. These results are supported by those found by Chauhan et al, (2015) with a content of 4.78 mg EC3G/ 100g for fresh fruit and 4.67 mg EC3G/100g after drying.Moreover, Pourghayoumi et al, (2017) in the course of their study on some Iranian fig varieties showed a great diversity in anthocyanin contents which ranged from 0.8 to 4.44 mg EC3G/ 100g.These data are almost similar to those found by Hoxha et al, (2015) who revealed anthocyanin contents for Albanian dry figs ranging from 0 to 5.32 mg EC3G / 100g.Among phenolic compounds of fig (Ficus carica) there are also flavonoids, which are pigments with an important role in the growth and defense of the plant (Pietta et al., 2003). It has been recognized that flavonoids have significant antioxidant activity and their effects on human nutrition and health are crucial (Bachir Bey and Louaileche, 2015). The estimation of the content of total flavonoids is carried out by a spectrophotometric method which is the method of the Aluminium Trichloride. Indeed, the flavonoids have a free OH group likely to give in the presence of aluminium chloride a yellowish complex by chelation of the Al^{+3} ion; the intensity of the yellow coloration is proportional to the quantity of flavonoids present in the extract. The absorbance is measured at 430 nm and the flavonoid content is determined by reference to a calibration curve performed with quercetin or catechin (standards). The results are then expressed in mg quercetin (EQ) or catechin (EC) equivalent /100g DM (Djeridane et al., 2006).Various studies have estimated the flavonoid content in methanolic extracts of dried figs. Bachir Bey and Louaileche, (2015) reported a flavonoid content of 126.55 mg EQ / 100g. These results are higher than those found by Ouchemoukh et al, (2012) who estimated the flavonoid content of some dried fruits marketed in the markets of Bejaïa, and proved that dried fig is the richest fruit in flavonoids (105.6 mg EQ /100g DM) compared to other dried fruits such as grapes and plums.It is known that during drying, the fruit is exposed to the sun and the skin, which accumulates high levels of antioxidants, will be the part most exposed to the sun. Therefore, the decrease in flavonoids after drying is not surprising, as these compounds act as UV filters, protecting certain cellular structures, such as chloroplasts, from the harmful effects of UV radiation (Treutter, 2006). These facts were confirmed by the work conducted by Manoj et al, (2018) and Kamiloglu and Capanoglu, (2015) where they detected a content of 0.21 mg EQ /100g for fresh figs and 0.19 mg EQ /100 g after drying, and a concentration of 66 mg EC /100g in fresh state and 52 mg EC /100 g in dry state for the first and second study, respectively.Flavonoid degradation does not only occur due to temperature and heating, it will additionally rely on alternative parameters such as pH, the presence of oxygen, and thus the presence of different phytochemicals in the medium (Ioannou et al., 2012).

IV.1.4. Influence of sun drying treatment on the activity antioxidant

Many diseases such as cancer, inflammation, atherosclerosis, Alzheimer's disease and Parkinson's disease are related to damage mediated by reactive oxygen species of biological macromolecules, which result from an imbalance between radical generating and radical scavenging systems (Taubert et al., 2003). For this, several researches also perform the evaluation of the antioxidant potential of polyphenolic extracts (Aljane, 2018).The evaluation of antioxidant activity can be performed by several different methods, among others: the DPPH (1,1-diphenyl-2-picrylhydrazil) test, the iron reducing power test (FRAP) and the ABTS (2, 2-azinobis (3- ethylbenzothiazoline)-6-sulfonic acid) radical trapping test.The DPPH test is widely used to determine the antiradical capacity of various samples. Indeed, when a DPPH solution is mixed with a substance that can donate a hydrogen atom, it results in the reduced form with a loss of the purple color. The discoloration will be directly proportional to the number of captured protons and can be followed by reading the absorbance of the reaction medium at 517 nm. It allows to evaluate the rate of reduction of DPPH and thus provides a convenient way to measure the antioxidant power of the studied extracts (Brand-Williams et al., 1995). The antiradical activity can be estimated according to the following equation:Antiradical activity (%) = [(Ac -At) / Ac] x 100

- **Ac**: Absorbance at 517 nm of the control.
- **At**: Absorbance at 517 nm of the tested extract.

The study conducted by Chauhan et al, (2015) showed that the antioxidant activity of dried figs was equivalent to 75.36%, this result was higher than those obtained by Pourghayoumi et al, (2017) and by Bachir Bey and Louaileche, (2015) who reported values equal to 51.37 % and 41.63%, respectively.Like the DPPH test, the determination of iron reducing power is a simple analysis in its application and is in every way complementary to the DPPH test. This method is based on the ability of extracts to reduce ferric iron (Fe^{3+}) to ferrous iron (Fe^{2+}).

The mechanism is known to be an indicator of electron donating activity, characteristic of the antioxidant activity of polyphenols (Yildrim et al., 2001). Indeed, the ferric-tripyridyltriazine complex is reduced to the ferrous-tripyridyltriazine form in the presence of antioxidants; the complex loses its yellow color to a dark blue. This coloringmeasured at 700 nm is proportional to the concentration of antioxidants present in the samples. In order to evaluate the antioxidant efficiency of some Turkish dried fig varieties, the iron reducing power (FRAP) method was adopted in different works including the one carried out by Nakilcioglu and Hisil, (2013) where they reported an average reducing activity of 222.71 mg FeSO4/100 g DM. This value is much lower than that reported in the study conducted by Reddy et al, (2010) on Indian fig varieties, where they revealed a content of 3578.69 mg FeSO4/100 g DM. In another study conducted by Capanoglu, (2014) on Turkish dried fruit varieties including fig, the FRAP reduction rate was equal to 117 mg ET/100g. This value is lower than that obtained by Kamiloglu and Capanoglu, (2015) who estimated a content of 140 mg ET/100g.The antioxidant activity of fig phenolic extracts can also be evaluated using the $ABTS^{-+}$ cation radical scavenging method. The latter is based on the ability of the compounds to trap the cationic radical $ABTS^{-+}$, (ammonium salt of 2,2'- azinobis (3-ethylbenzothiazoline)-6-sulfonic acid), which exhibits a visible absorption spectrum with three maxima at 645, 734, and 815 nm (Thaipong et al., 2006). Upon reaction with potassium persulfate (K2S2O8), ABTS forms the blue to green colored ABTS radical^{+} . The addition o f an antioxidant will reduce this radical and cause the decoloration of the mixture. The decoloration of the radical, measured by spectrophotometry at 734 nm is proportional to the concentration of antioxidants.Hoxha et al, (2015) and Hoxha and Kongoli, (2016) evaluated the antioxidant activity of two Albanian dried fig varieties, where they recorded a significant difference between their obtained results. The varieties grown in 2015 had an average value of 459.5 mol ascorbic acid equivalent /100g DM. A lower content for those grown in 2016 was reported (24.5 mol ascorbic acid equivalent /100g DM).While in another research performed by Konac et al., (2017) on Turkish fig varieties a value of 4.62 mmol ET /kg was estimated. Slightly higher results were reported in the study performed by Pellegrini et al., (2006) on Italian figs with a value of 5.02 mmol ET/ kg.According to the results previously cited in the different studies, the antioxidant capacity of figs was strongly correlated with the amount of phytochemicals. It should be noted that also that due to the different experimental conditions used in the aforementioned tests, no single test can fully characterize the profile of each sample. Therefore, their combination is recommended in order to obtain an overview of the antioxidant capacity of fresh and dried figs (Arvaniti et al., 2019).

IV.2. Drying in the oven

Dehydration of leafy tree fruits is normally achieved by sun-drying, but there are concerns about the safety of the final product, mainly because of the risk of aflatoxin development. These concerns can be overcome by artificial drying (oven dehydration) (Piga et al., 2004).Hot air dehydration has gained importance because it has many advantages over sun drying, including , better sanitary conditions brought about due to reduced contamination by dust and other foreign materials, drying parameters can be precisely set, controlled and changed throughout the processing time, resulting in a more uniformly uniform product with less quality degradation; also dehydration is not conditioned by rain or weather changes (Hoxha and Kongoli, 2016).

IV.2.1. Influence of treatment of drying at oven on the physicochemical characteristics

The effect of hot air drying on several quality parameters of figs (Ficus carica) has been initiated in some recent works especially in the one conducted by Khairuddin et al., (2017) on some types of Malaysian fruits including figs. A moisture content of 30.86% was determined after drying in this study.In a comparative study conducted by Eshak, (2018), the impact of two drying methods (sun and oven) on the quality attributes of some vegetables and fruits in Egypt (including figs) was investigated. A significant difference could be traced in the moisture content between fresh and dried samples. Oven-dried samples had a lower moisture content (3.10%) than sun-dried samples (6.96%). This difference is likely due to the high temperature of the oven dryer and the moderately short times compared to sun-drying (Fana et al., 2015).Regarding the total dry matter content of figs, Piga et al.evaluated this parameter in their study conducted on Italian figs, where they reported after oven drying an average content of 86.43%. These data are consistent with the recommended DM content for oven-dried fig (Hoxha and Kongoli, 2016).Both drying processes (sun and oven) have an increasing effect on the dry matter content, this fact was confirmed in the work done by Hoxha and Kongoli, (2016) where the initial dry matter varied from 30.19 to 32.08%, after sun drying it increased from 74.4 to 81.64%, while after the hot air dehydration process it increased from 81.22 to 83.31%.The estimation of titratable acidity and pH of oven-dried figs has been reported in few studies including that of Piga et al. Indeed, in fresh state the reported values were 5.67 and 0.87% for pH and acidity, respectively. The results obtained after drying decreased to a pH value equal to 4.94 and a titratable acidity equivalent to 0.48%.According to Hoxha and Kongoli, (2016) hot air dehydration resulted in an increase in titratable acidity (0.83-1.024%) compared to fresh fruit (0.38-0.62%), this is due to the fact that dried fruit samples contain less water, and organic acids are the most predominant. On the other hand, a decrease in pH was observed between fresh (4.76-4.67) and dried samples (4.35-4.19).Overall an inverse relationship was determined between pH and titratable acidity values for fresh and dried figs. The ash content of a food product is an index of nutritional value (mineral content, safety and quality). According to the results obtained by Eshak, (2018) the ash content of fresh samples (1.84%) was significantly lower than sun dried (5.69%) and oven dried (7.30%). Furthermore in the study conducted by Khairuddin et al, (2017) a lower ash content (3.79%) was determined after oven drying compared to the above. Overall, the samples showed a significant increase in ash content after drying due to the removal of water, thus increasing the nutrient concentration. The increase in ash content after drying can also be explained by the low volatility of minerals, which are not destroyed by heating (Eshak, 2018).

IV.2.2. Influence of oven drying treatment on the nutritional value

Among the most studied nutrient compounds in fig are distinguished; carbohydrates, proteins and fat. Regarding carbohydrates, a content of 57.36% in oven-dried figs was reported by Khairuddin et al, (2017). This value is lower than that reported by Eshak, (2018) who reported that oven dried fig samples recorded a lower value (73.7%) comparing to sun dried (76.9%).About the evaluation of protein content in fresh and dry figs Eshak, (2018) noted in these results that oven dried samples recorded the highest protein content (8.42%) comparing to sun dried (7.45%) and fresh samples (4.60%). On the other hand a lower protein content (3.93%) was reported by Khairuddin et al, (2017) in oven dried figs.The difference in results for figs dried by the two processes is due to the facts that sun-drying reduced the crude protein content compared to oven-drying due to denaturation of these heat-labile molecules and prolonged drying in an uncontrolled environment (Olapade and Ogunade, 2014).Although figs are not a good source of protein, they can contribute to the human diet with a high quality of some essential amino acids (Eshak, 2018).Fat (MG) contributes to the total energy content of a food product. Therefore, its value is important in estimating the caloric value of a food product (Kumur and Neeraj, 2018). Eshak, (2018) determined a value of 0.36% MG for fresh fig in his study. This value was lower compared to the oven dried sample (1.28%) which was itself higher than the value obtained for the sun dried sample (0.98%). According to the results obtained by Eshak, (2018) the low fat content of fig samples compared to other studied fruits and vegetables indicates their stability towards lypolitic rancidity reactions, hence better shelf stability (Nishant and Neeraj, 2018).

IV.2.3. Influence of oven drying treatment on total phenolic content and antioxidant activity

Dietary fruits including fig provide a reasonable amount of phenolic compounds that act as physiological antioxidants (Nakilcioglu and Hisil, 2013). Khairuddin et al, (2017) reported in their study that the methanolic extracts of Malaysian oven-dried figs contained about 151.04 mg EAG/100 g of total polyphenols. This result is lower than that reported by Hoxha and Kongoli, (2016) who conducted their study on two varieties of oven-dried Albanian figs, where they reported values ranging from 168.42 to 214.59 mg EAG/100 g. This last study also showed that the applied drying methods (sun and oven drying) have an influence on the total polyphenol content. On the other hand, lower results (59 mg EAG / 100g) were obtained from Slovenian figs studied by Slatnar et al, (2011).The antioxidant profile of hot air (oven) dried fig extracts has been addressed in few recent studies. In the studies that we will mention below, the evaluation of antioxidant activity was limited by the use of the DPPH and ABTS methods.Hoxha and Kongoli, (2016); Slatnar et al, (2011) adopted DPPH method for estimation of antioxidant activity in the different samples of sun dried and hot air (oven dried) figs. In the report of the study conducted by Hoxha and Kongoli, (2016) it was noted that the antioxidant activity of hot air dried samples (5-6.5 mol ET/100g) was higher than sun dried samples (3.9-5 mol ET/100g).Contradictory results were reported by Slatnar et al, (2011) where they found that sun-drying had a better influence on the antioxidant potential of figs compared to oven-drying with contents of 25 mg ascorbic acid equivalent /100g and 23 mg ascorbic acid equivalent /100g, respectively.Regarding the evaluation of antioxidant activity by ABTS method Konac et al, (2017) detected a content ranging from 387.26 to 674.55 µmol TE /100 g DM. While Hoxha and Kongoli, (2016) determined a content between 29 and 35 mol ascorbic acid equivalent /100 g.

IV.3. Drying by micro waves

The quality of the dried product depends on the drying conditions and methods to be used. For this, a faster and more efficient drying process, such as the microwave (MO) drying method should be considered for fruit dehydration. Recently, the microwave drying method has been proposed as an efficient drying alternative to conventional drying. MO leads to volumetric heating which means that all materials such as fruits and vegetables can be heated to the desired temperature at the same time (Parit and Prabhu,2017).Research on microwave drying has so far focused mainly on the fundamental aspects of industrial application. For this reason very few studies have been performed to evaluate the effect of microwave drying on the final quality of fig (Ficus carica). Therefore in this section we will only discuss a recent work done by Chauhan et al, (2015) on fig. The results of the analysis of the latter will be reinforced by those cited in work done on other MO dried fruits and vegetables.In order to raise awareness of the effects of processing methods on the nutritional value of fruits and their habit of consuming fresh seasonal fruits, Chauhan et al . , (2015) examined the effect of three drying treatments, including sun-drying and microwave technology, on the physicochemical, nutritional, phytochemical composition and especially antioxidant activity of fig (Ficus carica).In this study two physico-chemical parameters were evaluated, namely moisture and ash. It was observed that the moisture content reached after microwave drying (25.43%) was almost similar to that of sun drying (25.86%) and obviously lower than the result obtained in the fresh state (80.2%). Indeed, Mcloughlin et al (2003) indicated that microwave energy is rapidly absorbed by water molecules, resulting in rapid evaporation of water, leading to higher drying rates.For ash approximate results were obtained after microwave and sun drying with values of 4.30% and 4.42%, respectively. However, these contents were slightly higher than that obtained with fresh samples (4%).The nutritional composition of carbohydrates, proteins and fats was also determined by Chauhan et al., (2015). Regarding the carbohydrate content, it was noted that the microwave and sun dried samples showed similar results (65.18g/100g and 65.15g/100g, respectively) and obviously higher than the value obtained in the fresh state (16.3g /100g).Concerning the protein content of the MO dried figs, the results showed an average value (3.18%) slightly higher than that obtained during sun drying (3.01%) but higher compared to the fresh samples (0.53%).In the same study, it was found that the fat content was almost equivalent between microwave, sun-dried and fresh samples with values ranging from 0.53 to 0.59%.Phenolic content was also estimated by Chauhan et al, (2015), where they detected similar average values of about 5 mg EAT/100 g during microwave and sun drying. On the other hand, in a study performed by Arslan and Özcan,

(2010) on a Turkish onion variety (A. cepa), the phenolic content was evaluated after sun and microwave drying. It was observed that the total polyphenol content was higher in the MO dried samples (1624 mg/100 g) compared to that estimated during sun drying (472 mg/100 g).On the other hand, Chauhan et al, (2015) also studied the impact of drying processes on antioxidant activity through two tests, DPPH test and ABTS radical scavenging test. According to the results obtained, the averages of the DPPH inhibition percentage varied between 73.42%, 75.36%, and 75.84% for fresh, sun-dried, and microwave-dried samples, respectively. When analyzed by the ABTS method, the value recorded for the microwave-dried figs was higher (78.54%) than that obtained with the sun-dried (76.55%) and fresh samples (76.22%). These results are supported by those obtained by Valadez-Carmona et al, (2016) who studied the effect of microwave and oven drying on the polyphenolic composition and antioxidant activity of coconut (Cocos nucifera L.) husk.Overall, it was concluded that microwave drying was a better approach to drying coconut husk in terms of phenolic content and activity antioxidant compared to the hot air oven method, where they recorded a total phenol content of 45.5 mg/100g DM and a reducing activity assessed by ABTS equivalent to 1226.6 µmol TE/g after microwave drying, and lower values equal to 35.8 mg/100g DM for polyphenolic content and 743.8 µmol TE/g for ABTS reducing activity after hot air drying . In the same context Ozcan-Sinir et al, (2019) also evaluated the phenolic content and antioxidant activity of a kind of citrus fruit named "kumquat" from Turkey dried with MO and hot air. Higher values of total phenols (2800 mg EAG/100g DM) and antioxidant activity (9 µmol ET /g DM) were reported after MO drying, comparing to that obtained during hot air drying with values equal to 2300 mg EAG/100g DM and 5 µmol ET /g DM, respectively.

CONCLUSION

The fig (Ficus carica) is among the climacteric, seasonal fruits whose availability for prolonged consumption has been made possible by drying. The latter has proven to be a reliable preservation method for figs, in terms of technical feasibility and nutritional quality. In addition to their energy role, dried figs have good levels of nutrients and represent an important source of fiber, vitamins, minerals ... etc..

They also contain many bioactive compounds such as polyphenols, flavonoids, anthocyanins that characterize this fruit by remarkable antioxidant and therapeutic properties.Several studies have evaluated the impact of different drying processes such as sun drying, oven drying and microwave assisted drying on the physicochemical, nutritional, phytochemical quality and especially the antioxidant activity of the fig (Ficus carica).

The works that have dealt with the influence of sun drying on the quality attributes of the fig, including physicochemical characteristics, have shown that it causes the removal of water from the fruit leading to a significant decrease in moisture, thus influencing the appearance of the fig (length and width).From the results obtained from previous studies, we found that the nutrient composition is positively influenced by sun drying.

Total carbohydrates form the largest part of the fruit, which confirms that the dried fig is a high energy fruit. On the other hand, it is not considered as a source of fat.Regarding the phytochemical composition and particularly the content of total polyphenols, contradictory results have been reported by different authors, indicating that sun-drying can have a positive or negative effect on the content of these compounds.

Equally opposite results have been reported in the different studies dealing with the impact of this drying process on antioxidant activity.Oven drying has also been the subject of several recent studies.

According to the data obtained by several authors, oven drying has shown a better efficiencycompared to sun-drying with respect to physicochemical characteristics, namely moisture, dry matter and ash content. Regarding nutritional value, oven drying also showed a more positive effect than sun drying regarding protein and fat content. Also a lower carbohydrate content in oven-dried figs compared to sun-dried ones has been reported in the literature.Several studies have shown that the oven-dried fig represents a good source of polyphenols with values that can reach 214 mg EAG/100g, which has led to confirm its very important antioxidant power.Recently an alternative method of microwave assisted drying has been the subject of some studies.

According to the results obtained, microwave drying improves the physicochemical and nutritional quality, in particular the carbohydrate and protein content. It has also been proved that it has an increasing effect on the phytochemical composition and especially on the antioxidant activity.In general, all the work carried out has allowed us on the one hand, to deduce that the type of variety of figs and the nature of the drying process applied can have an influence on the final quality of the dried fig, and on the other hand, to confirm the advantage of the consumption of this dried fruit in view of its nutritional interest and its richness in antioxidants.Finally, as a perspective, we wish to carry out an experimental study on the different varieties of figs (Ficus carica) existing in our region "Jijel" in :

• Evaluating the impact of several drying processes (sun, oven and microwave) on the quality attributes of figs, namely physicochemical, nutritional and phytochemical quality, including antioxidant activity.

• We also perform sensory and microbiological analyses to determine the most efficient drying method.

• Contributing to shape a national database on the different varieties of dried figs existing in Algeria.

BIBLIOGRAPHIC REFERENCES

-A-

Abou-Farrag, H.T., Abdel-Nabey, A.A., Abou-Gharbia, H.A., & Osman, H.O.A. (2013).
Physicochemical and Technological Studies on Some Local Egyptian Varieties of Fig (Ficus
carica L.). Food Science and Technology ,34(2) ,190-203.

Abul-Fadl, M. M ., Ghanem, T.H., EL-Badry ,N.,& Nasr, A. (2015).Effect of some Different
Drying Methods on Quality Criteria of Dried Fig Fruits ,4(4),1-19 .

Aguilera, J. M., Chiralt, A., & Fito, P. (2003). Food dehydration and product structure. Trends
in Food Science & Technology, 14, 432-437.

Ait-Haddou, L., Blenzar, A., Messaoud, Z., Van Damme, P., Boutkhil, S., & Boukdame,
A.(2014) .The effect of cultivar, pretreatment and drying technique on some physico-chemical
parameters of dried figs from seven local cultivars of fig tree (Ficus Carica L.) in Morocco .
European Journal of Scientific Research,121 (4),336-346.

Aksoy, U. (1994). Present status and future prospects of underutilized fruit production in
Turkey. First Meeting CIHEAM Cooperative Research Network on Underutilized Fruit Trees.
Zaragoza, Spain, 84-94.

Al-Askari, G., Kahouadji, A., Khedid , K ., Charof, R., & Mennane, Z. (2012).
Physicochemical and microbiological characterizations of dry fig collected from the markets
of Rabat- Salé, Temara and Casablanca, 7 (26) ,12-19.

Albitar, N. (2010). Comparative study of drying processes coupled with texturing by
Controlled Instantaneous Release DIC, in terms of kinetics and nutritional quality.
Applications to the valorization of agro-industrial waste. PhD thesis: Industrial Process
Engineering. France: University of La Rochelle. 143p.

Ali-Rachedi, F., Meraghni, S., & Touaibia, N. (2018). Quantitative analysis of phenolic
compounds of an Algerian endemic Scabiosa Atropurpurea sub.Maritima L. Bulletin de la
Société Royale des Sciences de Liège, 87(1) ,13-21.

Aljan, F. (2018). Evaluation of phenolic compounds and antioxidant activities of figs (Ficus
carica L.). Cieham-Bari. [2nd] Mediterranean Forum for PhD students and young

Al-Maliki, A. D.M. (2012). Isolation and Identification of Tannins from Ficus carica L.
Leaves and Study of Their Medicinal Activity Against Pathogenic Bacteria. Thi-Qar Journal
of Science, 3 (3),2-11.

Al-Snafi, A. E. (2017). Nutritional and pharmacological importance of Ficus carica: A review.
Journal of Pharmacy and Biological Sciences, 7 (3), 33-48.

Arslan, D., & Özcan, M.M. (2010). Study the effect of sun, oven and microwave drying on quality of onion slices. Journal of food science and technology, 43(7), 1121-1127.

Arvaniti, O.S., Samarasa, Y., Gatidoub, G., Thomaidisc, N.S., & Stasinakisb, A.S. (2019). Review on fresh and dried figs: Chemical analysis and occurrence of phytochemical compounds, antioxidant capacity and health effects. Food Research International, 119(1), 244-267.

AOAC (1990). Official Methods of Analysis. [15th] edition The Association of Official Analytical Chemists.

AOAC (1997). Official Methods of Analysis. [18th] edition. The Association of Official Analytical Chemists.

AOAC (2000). Official Methods of Analysis. [17thedition], The Association of Official Analytical Chemists.

-B-

Babalis, S. J., Papanicolaou, E., Kyriakis, N., & Belessiotis, V. G. (2006). Evaluation of thinlayer drying models for describing drying kinetics of figs (Ficus carica). Journal of Food Engineering, 75 (2), 205-214.

Bachir Bey, M ., Louaileche, H., & Zemouri, S. (2013). Optimization of Phenolic Compound Recovery and Antioxidant: Activity of Light and Dark Dried Fig (Ficus carica L.) Varieties .The Food Science and Biotechnology, 22(6), 1613-1619.

Bachir Bey, M., & Louaileche, H. (2015). A comparative study of phytochemical profile and in vitro antioxidant activities of dark and light dried fig (Ficus carica L.) varieties. The Journal of Phytopharmacology,4 (1),41-48.

Bachir Bey, M., Louaileche, H., Meziant, L., Richard, G., & Fauconnier, M.L. (2016). Effects of sun-drying on physicochemical characteristics, phenolic composition and in vitro antioxidant activity of dark fig varieties. Journal of Food Processing and Preservation, 1-8.

Bauwens, P. (2008). Figs of all countries. Ed. Edisud, 96 p.
Belaid, D .(2015). Guide Du Sécheur De figues. In: Algeria - Manuel De Séchage Des Fruits.

Algeria ,65p.
Belitz, H. D., Grosch, W., & Schieberle, P. (2009). Fruits and Fruit Products. Food Chemistry.

Ed. Springer, 807-861.
Benettayeb, Z.E. (2018). Molecular and morphological characterization of the Algerian fig tree (Ficus carica L.). PhD thesis: Science of nature and life. Algeria: University of Oran, 98 p.
Benettayeb, Z. (1993).biology and ecology of fruit trees.Ed.OPU.Alger.140p.
Berg, C.C., & Wiebes, J.T. (1992). African fig trees and fig wasps. Royal Netherlands

Academy of Arts and Sciences. Papers of the Department of Physics, Royal Netherlands, 89 (2), 289 p.

Berthod, A., Billardello, B., & Geoffroy, S. (1999). Polyphenols in countercurrent chromatography. An example of large scale separation. Analusis, 27 (9), 750-757.

Boizot, N., & Charpentier, J.P. (2006). Rapid method for assessing the phenolic content of forest tree organs. Le Cahier des Techniques de l'INRA,79-81.
Brand-Williams, W., Cuvelier, M. E., & Berset, C. L. W. T. (1995). Use of a free radical method to evaluate antioxidant activity. LWT-Food science and Technology, 28(1), 25-30.

Brien, J., & Hardy, T.S. (2002). Fig growing in NSW. Agfact H3.1.19, [1st] edition Agdex 219.

Ed. Ann Munroe, 1-8.
Bruneton, J. (1987). Elements of phytochemistry and pharmacology. Ed. Lavoisier, 585 p.

-C-

Caliskan, O. (2015). Mediterranean figs (Ficus carica L.) Functional Food Properties. In: The mediterranean Diet: An Evidence-Based Approach.Ed.Elsevier, 629-627.

Çalişkan, O., & Polat, A. A. (2011). Phytochemical and antioxidant properties of selected fig (Ficus carica L.) accessions from the eastern Mediterranean region of Turkey. Scientia Horticulturae, 128(4), 473-478.

Cano-Chauca, M., Ramos, A. M., Stringheta, P. C., & Pereira, J. A. M. (2004). Drying curves and water activity evaluation of dried banana. Proceedings of the 14 [th] International Drying Symposium. Brazil: São Paulo.

Cantin, C.M., Palou, L., Bremer, V., Michailides, T.J., & Crisosto, C.H., (2011). Evaluation of the use of sulfur dioxide to reduce postharvest losses on dark and green figs. Postharvest Biology and Technology, 59 (2), 150-158.

Capanoglu, E. (2014). Investigating the Antioxidant Potential of Turkish Dried Fruits.

International. Journal of Food Properties, 17 (3), 690-702.
Carranza-Concha, J., Benlloch ,M., Camacho, M.M,.& Martínez-Navarrete ,N. (2012). Effects of drying and pretreatment on the nutritional and functional quality of grapes. Food and Bioproducts Processing, 90 (2), 243-248.

Castaneda-Ovando, A., Pacheco-Hernandez, M.D.L., Paez-Hernandez, M.E., Rodriguez, J.A., & Galan-Vidal, C.A. (2009). Chemicalstudies of anthocyanins: A review. Food Chemistry, 113,859-871.

Cendre, A. (2011). Innovative microwave fruit juice extraction process: manufacturing viability and nutritional quality of juices. PhD thesis: Food and Nutrition. France: the University of Avignon ,289 p.

Chahidi, B., El-Otmani, M., Jacquemond, C., Tijane, M., El-Mousadik, A., Srairi, I., & Luro F. (2008). Use of morphological, physiological traits and molecular markers for the assessment of genetic diversity in three clementine cultivars. Molecular Biology and Genetics, 331(1) ,1-12.

Chaker, S. (1997). Fig/Figuier. Berber Encyclopedia. Peeters Publishers, 18, 2825-2833.

Charles, J. (2002). Fabulous figs featured in California collection. Agricultural Research, 5, 14- 15.

Chauhan, A., Tanwar, B., & Intelli, A. (2015). Influence of Processing on Physicochemical, Nutritional and Phytochemical Composition of Ficus carica (Fig) Fruit. Journal of Pharmaceutical, Biological and Chemical Sciences, 8(6), 254-259.

Chawla, A., Kaur, R., & Sharma, A.K. (2012). Ficus carica L.: A review on its pharmacognostic, phytochemical and pharmacological aspects. International Journal of Pharmaceutical and Phyto-pharmacological Research, 1(4), 215-232.

Cheikh-Traoré, M. (2006). Study of the phytochemistry and biological activities of some plants used in the traditional treatment of dysmenorrhea in Mali. Doctoral thesis: Medicine of Pharmacy and Odonto-Stomatology. Mali : University of Bamako ,175p.

Chessa, I., & Nieddu, G. (2005). Analysis of diversity in the fruit tree genetic resources from a Mediterranean island. Genetic Resources and Crop Evolution, 52(3), 267-27.

Cheynier, V. (2005). Polyphenols in foods are more complex than often thought. The American Journal of Clinical Nutrition, 81(1), 223S-229S.

Clement, J. (1979). La santé par les plantes. Ed. Baudouin, Paris, 78-79.

Condit, I.J. (1955). Fig varieties: A Monograph Hilgardia. A Journal of Agricultural Science, 23(11), 323-539.

Cowan, M. M. (1999). Plant products as antimicrobial agents. Clinical microbiology reviews,12(4), 564-582.

Crisosto, C. H., Bremer, V., Ferguson, L., & Crisosto G. M. (2010). Evaluating quality attributes of four fresh fig (Ficus carica. L.) cultivars harvested at two maturity stages. Journal of Horticultural Science, 4 (45), 707-710.

Curtay, J. P., & Robin, J.M. (2000) Interest of antioxidant complexes. Nutrithérapie Info, 1- 4.

-D-

Debib, A., Tir-Touil, A., Mothana, R.A., Meddah, B., & Sonnet, P. (2013). Phenolic content, antioxidant and antimicrobial activities of two fruit samples of Algerian Ficus carica. L. Journal of Food Biochemistry, 38(2), 1-9.

Del Caro, A., & Piga, A. (2008). Polyphenol composition of peel and pulp of two Italian fresh fig fruit cultivars (Ficus carica L.). European Food Research Technology, 226(4), 715-719.

Delluc, L. (2004). Identification and functional characterization of two genes regulating the metabolism of phenolic compounds in grapes. PhD thesis: Life Sciences, Geosciences, Environmental Sciences. France: Bordeaux University, 310 p.

Dionne, J.Y. (2002). Carotenoids. Québec Pharmacies, 48 (9), 800-804.

Djeridane, A., Yousfi, M., Nadjemi, B., Boutassouna, D., Stocker, P., & Vidal, N. (2006). Antioxidants activities of some Algerian medicinal plants extract containing phenolic compounds. Food Chemistry, 97(4), 654-660.

Doymaz, I. (2005). Sun drying of figs: an experimental study. Journal of Food Engineering, 71 (4), 403-407.

Doymaz, I., Gorel, O., & Akgun, N.A. (2004). Drying Characteristics of the Solid By-product of Olive Oil Extraction. Biosystems Engineering, 88(2), 213-219.

Duenas, M., Pérez-Alonso, J.J., Santos-Buelga, C., & Escribano-Bailon, T. (2008). Anthocyanin composition in fig (Ficus carica L.). Journal of Food Composition and Analysis, 21 (2), 107-115.

-E-

El Bouzidi, S. (2002). The fig tree: history, ritual and symbolism in North Africa. Dialogues d'Histoire Ancienne, 28 (2), 103-120.

El Khaloui, M. (2010). Valorization of the fig in Morocco. Monthly information and liaison bulletin of the National Program of Technology Transfer in Agriculture, n° 186,1-

Eshak, N.S. (2018). Quality Attributes of Some Vegetables and Fruits Preserved by Sun and Oven Drying Methods. Alexandria Science Exchange Journal, 39 (4), 708-721.

European Pharmacopoeia Commission.(2007). 6th edition. Strasbourg: Council of Europe.

-F-

Faleh, E., Ghaffari, A., & Ferchichi, A.(2015).Polyphenol and soluble Sugars contents of Tunisian Dried Fig .Journal of new sciences, Agriculture and Biotechnology, 24(6), 1126-1129.

Faleh, E., Oliveira, A. P., Valentão, P., Ferchichi, A., Silva, B.M., & Andrade, P. B. (2012). Influence of Tunisian Ficus carica fruit variability in phenolic profiles and in vitro radical scavenging potential .Brazilian Journal of Pharmacognosy,22(6),1282-1289. .

Fana, H., Shimelis, A., & Abrehet, F. (2015). Effects of Pre-treatments and Drying Methods on Chemical Composition, Microbial and Sensory Quality of Orange-Fleshed Sweet Potato Flour and Porridge. American Journal of Food Science and Technology, 3 (3), 82-88.

FAO stat (2018). Recent FAO statistics in the field related to the fig sector.

Website: www.faostat.org. Consulted in April 2020.

Favier, J.C., Ireland-Ripert, J., Laussucq, C., & Feinberg, M. (1993). Répertoire général des aliments. Tome 3: table de composition des fruits exotiques, fruits de cueillette d'Afrique. Ed. ORSTOM and Tech & Doc, INRA, 31-34.

Feliachi, K. (2006). Second national report on the state of plant genetic resources for food and agriculture, Institut National de La Recherche Agronomique d'Algérie (INRA), Algiers, 67 p.

Ferradji, A., Malek, A., Bedoud, M., Baziz, R., & Aoua, S.A. (2001). Forced convection solar dryer for fruit drying in Algeria. Revue des Energies Renouvelables, 4(1) ,49-59.

Ferrão, T. S., Tischer, B., Menezes, M. F. S. C., Hecktheuer, L. H. R., Menezes, C. R., Barin, J. S., Michels, L., & Wagner, R. (2017). Effect of Microwave and Hot Air Drying on the Physicochemical Characteristics and Quality of Jelly Palm Pulp. Food Science and

Technology Research, 23 (6), 835-843.

Flaishman, M.A., Rover, V., & Stover, E. (2008). The fig: botany, horticulture, and breeding.

Horticultural Reviews, 34, 113-197.
Frutos, P., Hervás, G., Giráldez García, F., & Mantecón, A. (2004). Review. Tannins and ruminant nutrition. Spanish journal of agricultural research, 2 (2), 191-202.

-G-

Gamero, J.L. (2002). Production of fig trees: perspectives for the marketing of dried figs. In: Narjisse, H. Production of Figs: prospects for the marketing of dried figs. Morocco ,52-56.

Garrone, B. (1998). The fig tree. Ecologists of Euzière. 2nd edition. Presses du Midi, Montpelier, 111 p.

Giusti, M.M., & Wrolstad, R.E. (2001). Characterization and measurement of anthocyanins by UV-Visible Spectroscopy. Current Protocols in Food Analytical Chemistry, 1-13.

Goby, J. (2006). Fig tree cultivation in a continental climate. Association Saint Fiacre Loire Baratte, France, 1-7.

Golubev, V.N., Pilipenko, L.N., & Kakhniashvili, T.A. (1987). Fractionation and composition of the carbohydrates of Ficus carica. Plenum Publishing Corporation, 22, 631-634.

Gozlekci, S. (2011). Pomological traits of fig (Ficus carica L.) genotypes collected in the west mediterranean region in Turkey. The Journal of Animal & Plant Sciences, 21 (4), 646- 652.

Gramza, A., & Kolczak, J. (2005). Tea constituents (Camellia sinensis L.) as antioxidants in lipid systems. Trends in Food Science and Technology, 16(8), 351-358.

Guingard, J. (1996). Biochimie végétale. Ed. Lavoisier, Paris, 175-192 p.

Guvenc, M. E. (2009). Analysis of fatty acid and some lipophilic vitamins found in the fruits of the (Ficus carica) variety picked from the Adiyaman district, 4 (3), 320-323.

-H-

Haesslein, D., & Oreiller, S. (2008). Fresh or dried, the fig is revealed. Filière Nutrition et diététique. Haute école de santé Genève, 1-4.

Hiem, K. E., Tagliaferro, R. A., & Bobilya, J. D. (2002). Flavonoid antioxidants: Chemistry, metabolism and structure-activity relationships. Journal of Nutritional Biochemistry, 13, 572-584.

Hoxha, L., & Kongoli, R. (2016). Influence of drying process on phenolic content and antioxidant activity of two different autochthonous Albanian fig varieties. Scientific Papers. Series A. Agronomy, 510-514.

Hoxha, L., Kongoli, R., & Hoxha, M. (2015). Antioxidant activity of some dried autochthonous Albanian fig (Ficus carica L.) cultivars. International Journal of Crop Science and Technology, 1(2), 20-26.

-I-

Infanger, E. (2004). Swiss Nutritional Composition Table for Consumers. Swiss Society of Nutrition. Federal Institute of Technology. Federal Office of Public Health. Bern, 2nd edition, 138 p.

Institut National De La Recherche Agronomique d'Algérie (INRAA). (2006). Second national report on the status of plant genetic resources. Algeria, 92 p.

Technical Institute of Fruit Growing and Vine (ITAF). The culture of the fig tree.

Ioannou, I., Hafsa, I., Hamdi, S., Charbonnel , C., & Ghoul, M . (2012). Review of the effects of food processing and formulation on flavonols and anthocyanin behavior. Journal of Food Engineering, 111(2), 208-217.

IPGRI & CIHEAM. (2003). Descriptors for Fig. International Plant Genetic Resources Institute, Rome, Italy, and International Centre for Advanced Mediterranean Agronomic Studies, Paris, France.

-J-

Ife Fitz, J., & Bas, K . (2003). Fruit and vegetable preservation, Agrodok Series No. 3.

Bas Kuipers, 94 p.

Jeddi, L. (2009). Valorization of Taounate figs -Potential, mode and proposed strategies.

Report of the Provincial Directorate of Agriculture of Taounate, Morocco, 29 p.
Jiang , L., Shen, Z., Zheng, H., He, W., Deng ,G., & Lu, H. (2013). Non-invasive evaluation of fructose, glucose, and sucrose contents in fig fruits during development using chlorophyll fluorescence and chemometrics. Journal of Agricultural Science and Technology, 15 (2), 333-342.

Joseph, B., & Justin Raj, S. (2011). Pharmacognostic and phytochemical properties of Ficus carica Linn: An overview. International Journal of Pharmacy and Technology Research, 3(1), 08-12.

-K-

Kahrizi, D., Molsaghi, M., Faramarzi, A., Yari, k., Kazemi, E., Farhadzadeh, A.M., Hemati, S., Hozhabri, F., Asgari, H., Chaghamirza, k., Zebarjadi, A., &Yousofv, N. (2012). Medicinal Plant in Holy Quran. American Journal of Scientific Research, N°42, 62- 71.

Kakhniashvili, T.A., Kolesnik, A.A., Zherebin, Y.L,.& Golubev, V.N. (1987). Liposoluble Pigments of the Fruit of Ficus carica.Plenum Publishing Corporation, 477-479.

Kamiloglu, S., & Capanoglu, E. (2015). Polyphenol Content in Figs (Ficus carica L.): Effect of Sun-Drying . International Journal of Food Properties, 18(3), 521-535.

Kelm, A., Hammerstone, J. F., & Schmitz, H.H. (2005). Identification and quantitation of flavanols and proanthocyanidins in foods: How good are data?.Clinical and Developmental Immunology, 12 (1), 35-41.

Khairuddin ,M.F., Haron, H., Yahya, H ., & CheMalek, N.H. (2017).Nutrient Compositions and Total Polyphenol Contents of Selected Dried Fruits Available in Selangor, Malaysia .Journal of Agricultural Science, 9 (13),41-49.

Kjellberg, B, F., Doumesche, B., & Bronstein, J, L. (1987). Longevity of a fig wasp (Blastophaga psenes).Ecology, 91(2), 117-122.

Kolesnik, A. A., Kakhniashvili, T.A., Zherebin Yu, L., Golubev, V. N. & Pilipenko, L.N. (1987). Lipids of the fruit of Ficus carica. Chemistry of Natural Compounds, 22(4), 394-397.

Konak, R., Kosoglu, I., & Yemenıcıoglu, A. (2017). Effects of different drying methods on phenolic content antioxidant capacity and general characteristics of selected Turkish fig cultivars. Acta Horticulturae, 1173(58), 335-340.

-L-

Lacroix, M.(2008). Qualitative and quantitative variations in milk protein intake in animals and humans: metabolic implications. Doctoral thesis: Human nutrition. France: Institut des sciences et industries du vivant et de l'environnement, Agro Paris Tech.

Laleh, G. H., Frydoonfar, H., Heidary, R., Jameei, R., & Zare, S. (2006). The effect of light, temperature, pH and species on stability of anthocyanin pigments in four berberis species. Pakistan Journal of Nutrition, 5(1), 90-92.

Lansky, E.P., & Paavilainen, H.M. (2011). Figs: the genus Ficus.1 [st] edition.415p.

Lee, C.Y., Shallenberger, R. S., & Vittum, M. T. (1970). Free sugars in fruits and vegetables. Food Science and Technology, no. 1, 1-12.

Lien, E. J., Ren, S., Bui, H. H., & Wang, R. (1999).Quantitative structure-activity relationship analysis of phenolic antioxidants. Free Radical Biology and Medicine, 26 (3-4), 285- 294.

Lim, T.K. (2012). Edible medicinal and non-medicinal plants: Ficus carica. Moraceae Fruits. Ed. Springer Sciences Media B, 3,898 pp.

-M-

Mamouni, A. (2002). Caprification: Potentialities and constraints for the production of dried figs. In: Narjisse, H. Production de Figues: perspective pour la commercialisation des figues séches. Morocco, 42-51p.

Manoj, K., Bornare, D., & Kalyan, B. (2018). Effect of Drying on Physicochemical and Nutritional Quality of Ficus carica (Fig). International Journal of Engineering Research, 7(6), 117-121.

Marei, N., & Crane, J. C. (1971). Growth and respiratory response of fig (Ficus carica L. cv. Mission) Fruits to Ethylene. Plant Physiology, 48(3), 249-254.

Marfak, A. (2003). Gamma radiolysis of flavonoids, study of their reactivity with radicals from alcohols: formation of depsides. Doctoral thesis: Faculty of Medicine and Pharmacy. France: University of Limoges, 220p.

Martin, S., & Andriantsitohaina, L. (2002). Mechanism of cardiac and vascular protection of polyphenols at the endothelium. Annals of cardiology and angiology, 51(6), 304-315.

Martínez-García, J. J., Gallegos-Infante, J. A., Rocha-Guzmán, N. E., Ramírez-Baca, P., Candelas-Cadillo, M. G., & González-Laredo, R. F. (2013). Drying Parameters of Half- Cut and Ground Figs (Ficus carica L.) var. Mission and the Effect on Their Functional properties. Journal of Engineering, 1-8.

Mat Desa, W.N., Masita, M., & Fudholi, A. (2019). Review of drying technology of fig. Trends in Food Science & Technology, 88(1), 93-103.

Mauri, N. (1952). The cultivated fig trees in Algeria. Documents et renseignements agricoles, bulletin n°105, Alger.57 p.

Mcloughlin, C.M., McMinn, W.A.M., & Magee, T.R.A (2003). Microwave drying of multi component powder systems. Drying Technology, 21(2), 293-309.

Miguez Bemardez, M., De La Montana, J .M ., & Queijeiro ,G. J .(2004) .HPLC determination of sugars in varieties of chestnut fruits from Galicia (Spain). Journal of Food Composition and Analysis, 17, 63-67.

Morris, A., Barnett, A., & Burrows, O. (2004). Effect of processing on nutrient content of foods, 37(3), 160-164.

Mueller-Harvey, I., & Mc Allan, A.B., (1992). Tannins: their biochemistry and Nutritional properties . Advances in Plant Cell Biochemistry and Biotechnology,1, 151-217p.

-N-

Nakilcioglu, E., & Hısıl, Y. (2013) Research on the phenolic compounds in sarilop (Ficus carica L.) fig variety, 38 (5), 267-274.

Naikwadi, P. M., Chavan, U. D., Pawar, V.D., & Amarowicz, R. (2010). Studies on dehydration of figs using different sugar syrup Treatments. Journal of Food Science and Technology. 47(4), 442-445.

Naczk, M., & Shahidi, F. (2004). Extraction and analysis of phenolic in food. Journal of Chromatography A, 1054(1-2), 95-111. .

Nishant, K., & Neeraj. (2018). Study on physico-chemical and antioxidant properties of pomegranate peel. Journal of Pharmacognosy and Phytochemistry, 7(3), 2141-2147.

-O-

Okos, M. R. G., Narasimhan, R., Singh, K., Witnauer, A.C., & Campanella, O. (1992). Food dehydration. In: Heldman, D.R., Lund, D.B. (Eds.), Handbook of Food Engineering. Ed. Taylor and Francis Group, 2007, 1009p.

Olapade, A., & Ogunade, O. (2014). Production and evaluation of flours and crunchy snacks from sweet potato (Ipomeabatatas) and maize flours. International Food Research Journal, 21(1), 203-208.

Oliveira, A. P., Valentao, P., Pereira, J. A., Silva, B. M., Tavares, F., & Andrade, P. B.

(2009). Ficus carica L: Metabolic and biological screening. Food and Chemical Toxicology, 47 (11), 2841-2846.

Ouaouich, A., & Chimi, H. (2005). Guide du sécheur de figues. Small-scale agro-industrial entrepreneurship development project in peri-urban and rural areas of priority regions with a focus on women in Morocco, 1-27.

Ouchemoukh, S., Hachoud, S., Boudraham, H., Mokrani, A., & Louaileche, H. (2012). Antioxidant activities of some dried fruits consumed in Algeria. Food Science and Technology, 49 (2), 329-332.

Oukabli, A. (2003). The fig tree: a diversified genetic heritage to exploit. Programme National de Transfert de Technologie en Agriculture, Morocco, n°106 ,1-4.

Ozcan-Sinir, G., Ozkan-Karabacak, A., Tamer, C.E., & Utku Copur, O. (2019). The effect of hot air, vacuum and microwave drying on drying characteristics, rehydration capacity, color, total phenolic content and antioxidant capacity of Kumquat (Citrus japonica). Journal of Food Science and Technology, 39 (2), 475-484.

-P-

Pande, G., & Akoh, C.C. (2010). Organic acids, antioxidant capacity, phenolic content and lipid characterization of Georgia-grown underutilized fruit crops. Food Chemistry, 120(4), 1067-1075.

Parit, R.K., & Prabhu, C.S. (2017). Microwave Fruit and Vegetables Drying. International Advanced Research Journal in Science, Engineering and Technology, 4(2), 82-84.

Patil, V.V., & Patil V.R. (2011). Ficus carica Linn: An Overview. Research Journal of Medicinal Plant, 5 (3), 246-253.

Pellegrini, N., Serafini, M., Salvatore, S., Del Rio, D., Bianchi, M., & Brighenti, F. (2006).Total antioxidant capacity of spices, dried fruits, nuts, pulses, cereals and sweets consumed in Italy assessed by three different in vitro assays. Molecular Nutrition &Food Research, 50(11), 1030-1038.

Perez, C., Canal, J. R., & Torres, M. D. (2003). Experimental diabetes treated with Ficus carica. extract: effect on oxidative stress parameters. ActaDiabetol, 40(1), 3-8.

Peterson, J., & Dwyer, J. (1998). Flavonoids: Dietary occurrence and biochemical activity. Nutrition Research, 18 (12), 1995-2018.

Pietta, P., Gardana, C., & Pietta, A. (2003). Flavonoids in Herbs. In: Rice-Evans, C.A., & Packer, L. Flavonoids in Health and Disease. 2nd edition .New York: Marcel Dekker, 43-69.

Piga, A., Pinna, I., Ozer , K.B., Mario Agabbio, M., & Aksoy, U. (2004). Hot air dehydration of figs (Ficus carica L.): drying kinetics and quality loss. International Journal of Food Science and Technology, 39 (7), 793-799.

Pourghayoumi, M. R., Bakhshi, D., Rahemi, M., Noroozisharaf, A., Jafari, M ., Salehi ,M.,

Chamane, R ., & Hernandez, F .(2017). Phytochemical Attributes of Some Dried Fig.
(Ficus carica L.) Fruit Cultivars Grown in Iran, Agriculturae Conspectus Scientific, 81(3), 161-166.

-R-

Rebour, H. (1968). Mediterranean fruits other than citrus. Ed. La maison rustique, Paris, 330 p.Riberau-Gayon, P. (1968). Chemical properties of phenols: Application to natural products. In: Les composés phénoliques des végétaux. Ed. Dunod. 28-157.

Robards, K., & Antolovich, M. (1997). Analytical chemistry of fruit bioflavonoids. Analyst, 122, 11-34.

Reddy, C. V. K., Sreeramulu, D., & Raghunath, M. (2010). Antioxidant activity of fresh and dry fruits commonly consumed in India. Food Research International, 43(1), 285- 288.

-S-

Sen, F., Meyvaci, K. B., Turanli, F., & Aksoy, U. (2010). Effects of short-term controlled atmosphere treatment at elevated temperature on dried fig fruit. Journal of Stored Products Research, 46 (1), 28-33.

Sharifian, F., Modarres Motlagh, A., & Nikbakht, A.M. (2012). Pulsed microwave drying kinetics of fig fruit (Ficus carica L.). Australian Journal of Crop Science, 6 (10), 1441- 1447.

Sharma, M., Abid, R., & Sajgotra, M. (2017). Phytochemical Screening and Thin Layer Chromatography of Ficus carica Leaves Extract. UK Journal of Pharmaceutical and Biosciences, 5(1), 18-23.

Sirisha, S., Sreenivasulu, M., Sangeeta, K., & Chetty, C. M. (2010). Antioxidant properties of Ficus species: a review. International Journal of Pharmacy and Technical Research, 2 (4), 2174-2182.

Skiredj, A., Walali, L.D., & Ellatir, H. (2003). Almond, olive, fig, and pomegranate trees.
Transfer of technology in agriculture, Morocco, n°105, 1-4.

Slatnar, A., Klancar, U., Stampar, F., & Veberic, R .(2011). Effect of Drying of Figs (Ficus carica L.) on the Contents of Sugars, 2 Organic Acids, and Phenolic Compounds. Journal of Agricultural and Food Chemistry, 6-21.

Solomon, A., Golubowicz, S., Yablowicz, Z., Grossman, S., Bergman, M., Gottlieb, H.E., Altman, A., Kerem, Z., &Flaishman, M.A. (2006). Antioxidant Activities and Anthocyanin Content of Fresh Fruits of Common Fig (Ficus carica L.), Agricultural and Food Chemistry, 54(20), 7717-7723.

Soni, N., Mehta, S., Satpathy, G., & Gupta, R. K. (2014). Estimation of nutritional, phytochemical, antioxidant and antibacterial activity of dried fig (Ficus carica). Journal of Pharmacognosy and Phytochemistry, 3 (2), 158-165.

Spigno, G., Tramelli, L., & Faveri, D.M. (2007). Effects of extraction time, temperature and solvent on concentration and antioxidant activity of grape marc phenolics. Journal of Food Engineering, 81, 200-208.

Starr, F., Starr, K., & Loope, L. (2003). Ficus carica: Edible fig Moraceae. Biological Resources Division, 1-6.

Stover, E., Aradhya, M., Ferguson, L., & Crisosto, C.H. (2007). The Fig: Overview of an Ancient Fruit. Hort Science, 42(5), 1083-1087.

-T-

Tapiero, H., Townsend, D, M., & Tew, K, D. (2004). The role of carotenoids in the prevention of human pathologies. Biomedicine & Pharmacotherapy, 58(2), 100-110.

Taubert, D., Breitenbach, T., Lazar, A., Censarek, P., Harlfinger, S., & Berkels, R. (2003). Reaction rate constants of superoxide scavenging by plant antioxidants. Free Radical Biology and Medicine, 35, 1599-1607.

Thaipong, K., Boonprakob, U., Crosby, K., Cisneros-Zevallos, L., & Hawkink Byrne, D., (2006). Comparison of ABTS, DPPH, FRAP, and ORAC assays for estimating antioxidant activity from guava fruit extracts. Journal of Food Composition and Analysis, 19(6-7), 669-675.

Tous, J., & Ferguson, L. (1996). Mediterranean Fruits. In: Janick, J. Ed. Progress in New Crops, ASHS Press, Arlington, 416-430.

Trad, M., Bourvellec, C.L., & Gaaliche, B. (2013). Nutritional compounds in Figs from the Southern Mediterranean Region. International Journal of Food Properties, 17(3), 491-499.

Treutter, D. (2006). Significance of flavonoids in plant resistance: A review. Environmental Chemistry Letter, 4(3), 147-157.

Trifunschi, S.I., Munteanu, M.F.F., Ardelean, D.G., Orodani, M., OSSER, G.M., & Gligor,

R.I. (2015). Flavonoids and polyphenols content and antioxidant activity of Ficus carica L. extracts from Romania. Journal for Natural Science, no. 128, 57-65.

-V-

Valadez-Carmona, L., Cortez-Garcia, R. M., Plazola-Jacinto, C.P., Necoechea-Mondragon, H., & Ortiz-Moreno, A. (2016). Effect of microwave drying and oven drying on the water activity, color, phenolic compounds content and antioxidant activity of coconut husk

(Cocosnucifera L.). Journal of Food Science and Technology, 53(9), 3495-3501.

Vidaud, J. (1997). Le figuier. Paris: Centre technique interprofessionnel des fruits et légumes. Edi. SUDOC. 263 p.

Veberic, R., Colaric, M., & Stampar, F. (2008). Phenolic acids and flavonoids of fig (Ficus caria L.) in the northern Mediterranean region. Food Chemistry, 106(1), 153-157.

Vinson, J.A. (1999). The functional food properties of figs. American Association of Cereal Chemists, 44 (2), 82-87.

Vinson, J. A., Zubik, L., Bose, P., Samman, N., & Proch, J. (2005). Dried Fruits: Excellent in vitro and in vivo Antioxidants. Journal of the American College of Nutrition, 24(1), 44-50.

-W-

Wang, H., Cao, G., & Prior, R.L. (1997). Oxygen radical absorbing capacity of anthocyanins. Journal of Agriculture and Food Chemistry, 45(2), 304-309.
Weibes, J.T. (1979). Co-evolution of figs and their insect pollinators. Annual Review of Ecology and Systematics, 10(1), 1-12.

Wrolstad, R.E., Durst, R.W., & Lee, J. (2005). Tracking color and pigment changes in anthocyanin products. Food Science and Technology, 16(9), 423-428.

-Y-

Yancheva, S.V., Golubowicz, S., Yablowicz, Z., Perl, A., & Flaishman, M.A. (2005).Efficient Agrobacterium-mediated transformation and recovery of transgenic fig (Ficus carica L.) plants. Plant Science, 168(6), 1433-1441.

Yildrim, A., Mavi, A., & Kara, A. A. (2001). Determination of antioxidant and antimicrobial activities of Rumexcrispus L. extracts. Journal of agricultural and food chemistry, 49(8), 4083-4089.

-Z-

Zimmer, N., & Cordesse, R. (1996). Influence of tannins on the nutritional value of ruminant feeds. Institut national de la recherche agronomique Productions Animales, 9(3), 167-179.

Zoughlache, S. (2009). Study of the biological activity of extracts from the fruit of (Zizyphus lotus.L). Thesis of Magister. Batna: University-El hadj Lakhdar -Batna, 91p.

SUMMARY

The fig has been a typical fruit component of the Mediterranean diet for a very long time. Since it is a seasonal fruit with a perishable aspect, it must be preserved in its dry form which can be obtained by several drying processes namely sun, oven and microwave drying. Several investigations were interested in the study of the influence of these drying processes on the quality attributes of this fruit whether it is physicochemical, nutritional, phytochemical and especially on the antioxidant activity. The dried fig proved to be a fruit rich in nutrients as well as bioactive substances such as total polyphenols, flavonoids and anthocyanins which contribute to its very important antioxidant power. The results of the various works have shown that the drying techniques mentioned above have improved the physicochemical quality of the fig, on the one hand they have significantly decreased the moisture content (80.2 to 25.86% after sun drying), and on the other hand they have increased the content of certain parameters including ash (1.84% to 7.30% after oven drying). Likewise, it was indicated that these drying processes have a positive effect on the content of different nutritional and phytochemical compounds, namely carbohydrates (16.3 to 65.15 g/100g after sun drying), proteins (0.53 to 3.18% after microwave drying), dietary fiber (2.9 to 9.8 g/100g after sun drying), and also total polyphenols which reached a value of 151.04 mg EAG/100g after oven drying. Regarding the antioxidant activity, the ABTS assay showed a higher reducing activity in microwave dried figs (78.54%) than in fresh figs (76.22%). On the other hand, contradictory results were reported about the influence of sun and oven drying on the phytochemical profile and antioxidant activity, respectively. This work, opens new perspectives for the evaluation of the impact of these different drying processes on Algerian figs and in particular those of our region "Jijel".

Key word: dried fig, sun drying, oven drying, microwave drying, physicochemical quality, nutritional quality, phytochemical quality, antioxidant activity.

More
Books!

info@omniscriptum.com
www.omniscriptum.com
OMNIScriptum

Printed by Books on Demand GmbH, Norderstedt / Germany